OCEANOGRAPHY AND OCEAN ENGINEERING

DEEP SEA

BIODIVERSITY, HUMAN DIMENSION AND ECOLOGICAL SIGNIFICANCE

OCEANOGRAPHY AND OCEAN ENGINEERING

Additional books in this series can be found on Nova's website under the Series tab.

Additional e-books in this series can be found on Nova's website under the e-book tab.

OCEANOGRAPHY AND OCEAN ENGINEERING

DEEP SEA

BIODIVERSITY, HUMAN DIMENSION AND ECOLOGICAL SIGNIFICANCE

NIGEL T. WILSON
EDITOR

nova publishers
New York

For permission to use material from this book please contact us:
Telephone 631-231-7269; Fax 631-231-8175
Web Site: http://www.novapublishers.com

Additional color graphics may be available in the e-book version of this book.

Library of Congress Cataloging-in-Publication Data

ISBN: 978-1-63321-637-2

Published by Nova Science Publishers, Inc. † New York

CONTENTS

PREFACE

Approximately 70% of the surface of Earth is the bottom of the seafloor and about 80% of the seafloor is the deep seafloor (at deeper than 200 m below the sea-surface). This book discusses the deep sea. It provides topics on the biodiversity, human dimension and ecological significance of the deep sea, including the metagenomic and metatranscriptomic analysis discoveries; the biodiversity and life-history strategies of deep sea megafauna in the Mediterranean Sea; and prokaryotes in metal deposits on the deep seafloor.

Chapter 1 – The water body underlying the photic zone in oceans represents the largest water mass on earth (comprising 1.3×10^{18} m^3), and is also the largest aqueous habitat for various microorganisms. This realm differs distinctly from the photic zone, in terms of its relatively lower temperature (approximately 2~4°C), higher pressure and richer inorganic nutrients. Differences in physical geochemical parameters between upper-sea and deep-sea environments create fundamentally different challenges to microbial communities living in these environments. Recent studies found that prokaryotic microbes in deep-sea environments are well-adapted to the special dwelling environments after long evolution, carrying genetic features that enable them to live and reproduce in the extreme environmental conditions. Recent progress in sequencing technologies is fueling a rapid increase in the number and scope of deep-sea microbial community-targeted studies. While metagenomic analysis can provide information on the taxonomic composition and metabolic potential of microbial communities in the deep sea, metatranscriptomics serves to unveil the actual metabolic activities of the communities at a specific time and location, and how those activities are changing in response to environmental and biotic challenges. Here author provide a summary of recent progress in applying integrated metagenomic and metatranscriptomic analyses to uncover the special genetic features in the well-adapted deep-sea microbial communities.

Chapter 2 – The Mediterranean Sea is a relatively small but deep basin of 1450 m of mean depth situated between 30° N and 46° N and 5.5° W and 36° E. The only natural connection with the Atlantic Ocean is the Strait of Gibraltar, a narrow and shallow channel of 14 km width and 320 m depth. Given that the high evaporation rates are not compensated by inputs from rainfall and river discharges, the Mediterranean acts as a concentration basin in which water circulation is dominated mainly by salinity and temperature differences (thermohaline forcing). The Mediterranean has been compared with a miniature ocean for studies on oceanographic and biological processes taking place at larger scales. Although representing only 0.8% of the area and less than 0.25% of the volume of the world's oceans, the Mediterranean includes about 6% of known world marine fauna and 18% of world marine flora of which 28% are endemic. A total of 10000 to 12000 marine species (with 8500 species of macroscopic fauna) have been recorded. Despite its small dimensions, the Mediterranean coastal waters are a hot spot of biodiversity hosting a high percentage of endemic species. However, the diversity of deep megafauna is comparatively low, most probably due to the Gibraltar sill, which would act as a physical barrier for the entrance of larvae and deep-sea benthic organisms from the Atlantic. Although with some exceptions, there is a general decreasing pattern of biodiversity with increasing depth at different extent depending on the taxonomic group. Most megafaunal community studies have focused on fishes, decapod crustaceans and cephalopods and restricted to depths above 800 m. The epibenthic megafaunal communities below 200 m are still fairly unknown, yet cold-water corals communities have been described in recent years. Besides the environmental conditions typical of deep waters, Mediterranean deep-sea organisms must face high homeothermy (12.5-15.5°C), high salinity (38.0-39.5) and high oligotrophy (the eastern basin is one of the most oligotrophic areas of the world). The Mediterranean is among the most impacted seas in the world (e.g. crowding of coastal zones, biodiversity loss, overfishing and pollution). Climate change has been reported to affect deep-sea ecosystems and may have deleterious effects on the biodiversity of such sensitive habitats. In this chapter, author firstly introduce the Mediterranean Sea and describe its main oceanographic conditions. Secondly, author review current knowledge on the biodiversity and life-history strategies of Mediterranean deep-sea megafaunal groups such as fishes, crustaceans, cephalopods and cold-water corals. Finally, author discuss the main threats posed by human impacts on Mediterranean deep-sea ecosystems.

Chapter 3 – Approximately 70% of the surface of Earth is the bottom of the ocean (seabed or seafloor), about 80% of the seafloor is the deep seafloor (at deeper than 200 m below the sea-surface). Metal deposits accumulated on the seafloor have been found in various areas. Sulfide deposits containing metals such as Fe, Cu and Zn, called chimneys or mounds, and iron-silica-rich deposits occur in deep-sea hydrothermal fields. Ferromanganese deposits rich in Co, Cu, Ni and Zn, in addition to Fe and Mn, are found on the deep seafloor of both hydrothermal and non-hydrothermal fields. Previous studies have revealed that diverse prokaryotes (Bacteria and Archaea) are present on and within the seafloor metal deposits. Some species of prokaryotes can utilize Fe, Mn and S in respiration and fix inorganic carbon producing organic carbon. Others can accelerate dissolution or precipitation of metals. Metals such as Fe and Zn are essential for growth of almost all living organisms. Considering the wide distribution of the metal deposits on the deep seafloor and the amount of metals concentrated in the deposits, the activities of prokaryotes in the metal deposits potentially contribute to elemental cycling and maintenance of ecosystems in the global ocean. In this chapter, the recent studies of diversity and composition of prokaryotic communities in a variety of metal deposits on the deep seafloor are reviewed. Furthermore, the reported gene sequence data of the prokaryotes were re-analyzed using the latest bioinformatics tools. This provides novel insights into the characteristics of prokaryotic communities on the deep seafloor metal deposits.

In: Deep Sea
Editor: Nigel T. Wilson

ISBN: 978-1-63321-637-2

Chapter 1

METAGENOMIC AND METATRANSCRIPTOMIC ANALYSIS REVEAL GENETIC ADAPTATION OF DEEP-SEA MICROBIAL COMMUNITIES

Jieying Wu, Weimin Gao, Weiwen Zhang* and Deirdre R. Meldrum

Center for Biosignatures Discovery Automation, the Biodesign Institute, Arizona State University, Tempe, AZ, US

ABSTRACT

The water body underlying the photic zone in oceans represents the largest water mass on earth (comprising 1.3×10^{18} m^3), and is also the largest aqueous habitat for various microorganisms. This realm differs distinctly from the photic zone, in terms of its relatively lower temperature (approximately 2~4°C), higher pressure and richer inorganic nutrients. Differences in physical geochemical parameters between upper-sea and deep-sea environments create fundamentally different challenges to microbial communities living in these environments. Recent studies found that prokaryotic microbes in deep-sea environments are well-adapted to the special dwelling environments after long evolution,

* Email: Jieying.Wu@asu.edu; Current address: School of Chemical Engineering and Technology, Tianjin University, Tianjin 300072, P.R. China.

carrying genetic features that enable them to live and reproduce in the extreme environmental conditions. Recent progress in sequencing technologies is fueling a rapid increase in the number and scope of deep-sea microbial community-targeted studies. While metagenomic analysis can provide information on the taxonomic composition and metabolic potential of microbial communities in the deep sea, metatranscriptomics serves to unveil the actual metabolic activities of the communities at a specific time and location, and how those activities are changing in response to environmental and biotic challenges. Here we provide a summary of recent progress in applying integrated metagenomic and metatranscriptomic analyses to uncover the special genetic features in the well-adapted deep-sea microbial communities.

INTRODUCTION

The water body underlying the photic zone in the world's oceans (depths > 200 m), representing the largest water mass on earth (comprising 1.3×10^{18} m^3), is the largest aqueous habitat for microbial life (Orcutt et al., 2011). This fraction of the ocean is believed to harbor 6.5×10^{28} cells which accounts for about 55% of all the prokaryotes found in aquatic habitats (Whitman et al., 1998). A depth of 200 m is an average division between the deep ocean and the ocean surface. Sunlight can hardly penetrate beyond this depth, which causes the decrease of the light-dependent primary production. The realm of the ocean below 200 m is also different distinctly from the photic zone, presenting low temperature (approximately 2~4°C), higher pressure (pressure increases by roughly1 atm with each 10 m of depth) and higher inorganic nutrient levels (Aristegui et al., 2009). Inorganic compounds (NO^{3+}, PO_4^{3-}, Fe^{3+}, etc.) rarely exist in the photic zone, while they are re-mineralized and released into the water by microbial activity in the aphotic zone which makes the concentrations of these compounds increase in the depths below 200 m (Orcutt et al., 2011). In the middle depths of the ocean, the oxygen concentration is typically lower than in the surface water, mainly due to the decomposition of sinking organic material by aerobic respiration as well as the introduction of high-latitude, oxygen-rich surface water to deeper depths (Wyrtki 1962; Ulloa et al., 2012). However, oxygen is in general relatively abundant for the microbial population living in those depths, and oxygen-direct respiration is often observed in the microorganisms around this depth (Orcutt et al., 2011).

Differences in physical geochemical parameters between the upper level of sea water and deep sea suggest that microbial communities in these environments are confronted by fundamentally different challenges. Our knowledge of microbial processes below the sunlit ocean surface has been enormously expanded in the last 30 years, especially owing to the stirring discoveries in hydrothermal vents (D Prieur et al., 1995; Schrenk et al., 2003; Anantharaman et al., 2013), cold seeps (Fang et al., 2006; Orcutt et al., 2008), and whale falls (Deming et al., 1997; Smith 2003). These discoveries, as well as numerous ongoing studies, contribute to unveiling knowledge about microorganisms living in the mostly inaccessible deep, dark ocean, generate continuous and paradigm-shifting findings, and fundamentally change our thoughts toward deep-sea microbial communities (Orcutt et al., 2011). Studies from the Global Ocean Sampling (GOS) expedition indicated that new microorganism families are being discovered at a rate that is linear or almost linear with the addition of new sequences, implying that we are still far from discovering all protein families in nature. It was also predicted that marine microorganisms will continue to be a source of novel enzymes in the near future (Rusch et al., 2007).

Many marine microorganisms can synthesize various metabolic compounds. In recent decades (since the 1970s), a considerable amount of drug candidates were discovered from marine natural products (Haefner 2003). These natural products are a rich source of new chemical diversity and also a vital component of the pharmaceutical industry (Fenical & Jensen 2006). Hence, the development of tools to access deep-sea microorganisms and microbial community promises to provide insight into this significant new source of drug discovery and development. In this chapter, we introduce two leading approaches of studying microbial communities, metagenomics and metatranscriptomics, and their applications in exploring deep-sea microbial communities. Unique features and genetic adaptations of microbial communities living in this special environment are also discussed.

METAGENOMICS: A TOOL TO CATALOG MARINE MICROBIAL COMMUNITIES

The application of many recent new biotechnologies is fueling the exploration of microbial resources in ocean. Although many studies have led to cultivation, characterization and application of various ocean microorganisms, it is now widely accepted that the overwhelming majority (*i.e.* 99%) of the microorganisms in the ocean are uncultivable and thus not accessible by traditional microbiological methodologies (Rappé et al., 2002). To address the issues, various culture-independent approaches have been proposed to explore microbial communities in the ocean (Schmidt et al., 1991; Poretsky et al., 2005; Parro et al., 2007). These techniques allow analysis of microbial communities without the necessity to cultivate the microorganisms, and have greatly enlarged the fraction of microorganisms that can be identified (Fuhrman 2002). The traditional culture-independent techniques included cloning and sequencing of taxonomically indexed genes (Pace 1997), denaturant gradient gel electrophoresis (DGGE) (Muyzer & Smalla 1998), fluorescent *in situ* hybridization (FISH) (Amann et al., 1995), qPCR (Head et al., 1998), and DNA microarray (Giovannoni et al., 1990). Recent progress led to some new techniques such as 16S ribosomal RNA (rRNA) targeted sequencing which has offered a glimpse into the phylogenetic diversity of uncultured organisms (Tringe et al., 2005), and entered a new era based on next generation sequencing. Among the methods designed to study phylogenetics of uncultured organisms, metagenomics,--the genomic analysis of a population of microorganisms – has emerged as a powerful centerpiece (Handelsman 2004).

Since the first proposal by Pace et al., that microorganisms could be studied by extracting the mixed microbial nucleic acid from environmental samples and then analyzing DNA sequences (Pace et al., 1986), the term 'metagenomics' as a game-changing concept opened a new era in environmental microbial exploration. Metagenomics is defined as the study of the direct-isolated genetic components of a given environmental sample (Kennedy et al., 2010). The analysis of metagenomics can be either sequence driven or function driven (Handelsman 2004; Kennedy et al., 2010). Initially the technology involves environmental DNA isolation, DNA fragment cloning and transforming, followed by screening and sequencing of the transformants (Handelsman 2004). However, this technology is low in efficiency, very labor intensive, and it is difficult to achieve good coverage for genetic material isolated from the environmental microbial communities. Recent development in high-throughput and more affordable sequencing platforms have the advantages of high efficiency, labor savings, and high coverage. To date, several commercially available high-throughput sequencing platforms have been introduced, including Sanger sequencing using the chain termination method (Orsi et al., 2011; Youssef et al., 2012), 454 pyrosequencing (Rusch et al., 2007; Eloe et al., 2011; Wu et al., 2011), Illumina sequencing using synthesis sequencing technology (Shakya et al., 2013), SOLiD (Sequencing by Oligonucleotide Ligation and Detection) sequencing by ligation (Valouev et al., 2008), and Ion Torrent sequencing using ion semiconductor technology (Whiteley et al., 2012). These technologies can generate 1~5 million reads in a single run, with read lengths ranging from 50 ~ 1,000 bps. In recent years, these high-throughput sequencing technologies have fueled a rapid increase in the number and scope of microbial community-targeted studies of marine environment using metagenomic approaches. Genes involved in photosynthesis, carbon fixation, and nitrogen metabolism were found to be highly represented in the cDNA pool of surface water microbial community (Frias-Lopez et al., 2008). Comparative genomic analyses of stratified planktonic microbial communities in the ocean's interior revealed a depth-dependent trend of microbial community function in the North Pacific Subtropical Gyre. Genes involved in carbon and energy metabolism, attachment and motility, gene mobility, and host-viral interactions were found to be more enriched in the photic surface water (0~200 m) (DeLong et al., 2006). Comparative metagenomic analysis showed that the deep-sea microbial communities cope with its dwelling environment by a collection of very subtle adaptations, instead of dramatic alterations of gene content or structure. The studies found that deep-sea prokaryotes own higher metabolic versatility and genomic plasticity, and they typically have larger genome sizes and higher content of transposases and prophages (Konstantinidis et al., 2009).

As mentioned before, there are two approaches for metagenomic analysis, sequence-based and function-based (Handelsman 2004). Sequence-based metagenomic approaches compare sequence data with a previously constructed database, while function-based metagenomics identify genes directly by their function. The sequence-based analysis involves sequencing a large amount of random clones togenerate insightful data for samples harboring innumerous types of microorganisms(*e.g.* deep-sea microbial community) and provide information about genomic component and features, distribution and redundancy of functions in the community, and horizontal gene transfer (Venter et al., 2004; Handelsman 2004). On the other hand, for function-based metagenomics there is no need to recognize the targeted genes by sequences so this kind of approach can play a role in novel function detection (Handelsman 2004; Kennedy et al., 2010). The significant limitation of function-based metagenomics is that function is always linked with the 'expressed' genes, which imposes a big burden to obtain complete information of the target. To date, DNA sequence-based metagenomics that involves random community DNA sequencing, taxonomic and functional gene diversity determination, has become a standard tool for the analysis of natural microbial communities in various natural environments including alkaline oil sands tailings ponds (Saidi-Mehrabad et al., 2012), acidic hot springs (Jiménez et al., 2012), wastewater treatment bioreactors (Ye et al., 2012), Antarctic soil (Pearce et al., 2012), and oxygen minimum zones in the ocean (Stewart et al., 2011; Cassman et al., 2012).

METATRANSCRIPTOME: INSIGHTS INTO 'REAL-TIME' ACTIVITY

Although metagenomics is able to provide diversity information and functional potential of various samples including those of uncultured and unknown species (Ogura et al., 2011), it cannot provide information regarding expression level of functional genes and thus physiology of microbes under specific environmental conditions. To address the issue, metatranscriptomics was introduced into this field. The major metatranscriptomics technologies include microarray (Parro et al., 2007; Abell et al., 2012) and mRNA-derived cDNA clone libraries (Poretsky et al., 2005). A microarray is a tool for highly parallel hybridization of a single target to a multitude of probes in order to measure the expression levels of a large numbers of genes simultaneously or to genotype multiple regions of a genome (Schena et al., 1995; Lashkari et al., 1997), while the mRNA-derived cDNA clone library method is to apply randomly primed reverse transcription of enriched environmental mRNA to produce a cDNA template population and then to generate cDNA clone libraries (Poretsky et al., 2005). These technologies have been widely used and brought significant insights into the functionality of microbial communities. However, microarray analysis is limited to the known genes since it involves designing specific probes, while mRNA-derived cDNA clone library method may bring in additional biases, such as the relative abundance skewing of sequences because of the toxic products encoded by cloned transcripts (Gilbert et al., 2008). With new next-generation sequencing technologies, the current metatranscriptomics is similar to metagenomics except that it analyzes RNA samples. To date, it is possible to identify genes being transcribed in given conditions and further to depict gene expression patterns in natural microbial communities without any *prior* information (Gifford et al., 2011). Next-generation sequencing-based metatranscriptomics has allowed assessment of microbial functionality in the natural environments (Shi et al., 2009). The technology can overcome various constraints inherent in the use of traditional qPCR and microarrays to monitor gene expression in complex microbial communities, such as examination of a limited number of genes, a *prior* need to select gene targets, and requirements for probes or primers (Gilbert et al., 2008). In addition, sequencing-based metatranscriptomics has no significant bias towards any known sequences within microbial communities, and thus provides a new and powerful approach for identifying previously unknown genes and analyzing community-specific variants of functional genes in the nature (McGrath et al., 2008). In recent years, several metatranscriptomics analyses of marine and soil samples have been reported. The study made by Gilbert et al., (2008) provided a good example that when paired with metagenomic analysis, metatranscriptomic study can offer unprecedented exploration of both structural and functional information of a surface-ocean microbial community. In another study of gene expression in ocean microbial communities, half of the genes detected were undetected in previous DNA-based metagenomic surveys (Frias-Lopez et al., 2008). The study also showed that genes involved in photosynthesis, carbon fixation, and nitrogen acquisition occupy the largest portion in the cDNA pool (Frias-Lopez et al., 2008). While most of this work started from selective mRNA enrichment by depleting ribosomal RNA or enriching for polyA-tailed mRNA, in a recent study, a total RNA pool of microbial community, including both mRNA and rRNA, was analyzed to reveal both functionality and taxonomic diversity of the environments (Urich et al., 2008). This approach avoided extensive purification steps of mRNA, and enabled simple and quick preparation even from difficult samples. Together these studies have significantly enhanced our understanding of the diversity and functionality of microbial communities in the nature.

Defining the relationship between microbial community composition and function (metabolic characteristics) has been a major challenge in studying heterotrophic carbon cycling in marine systems (Teske et al., 2011). The combination of metagenomic and metatranscriptomic approaches has proven to be a powerful tool in deciphering both the phylogenetic composition, and the metabolic activities of deep-sea microbial communities. In a recent study, coupled metagenomic and metatranscriptomic analyses were utilized for taxonomic and functional characterization of marine microbial communities living at depths between 25-500 m below sea level, and have found differential relative transcriptional activities per cell (Shi et al., 2011). Functional characterization of highly expressed genes in this study led to the finding of taxon-specific contributions to specific biogeochemical processes, such as

Roseobacter sp. to aerobic anoxygenic phototrophy at 75 m, *Crenarchaea* to ammonia oxidation at 125 m, and transcriptionally active light-prefer *Prochlorococcus* population in the bottom of the photic zone (Shi et al., 2011). The results provided novel insights into not only microbial diversity but also specific metabolic processes transpiring in the ecosystems.

Technical Challenges in Applying Metatranscriptomics

Limitations in sample quality and quantity have become big obstacles for applying metagenomic/metatranscriptomic technologies to explore deep-sea microbial communities (Wu et al., 2011). Compared with biomass abundant samples, such as surface seawater, the intact nucleic acid that can be isolated from these environments is very limited in quantity, usually low to several hundred nanograms level. Moreover, environmental samples, different from pure culture grown in the laboratory, often contain various minerals and humic acids, which may inhibit amplification reactions (Broemeling et al., 2008; Demeke & Jenkins 2010). Thus, in most of these cases, quality of the isolated environmental nucleic acid is not comparable with that from laboratory grown pure culture either. Moreover, typical commercial next-generation sequencing platforms require 3-5 μg or more of input DNA/cDNA to produce reliable sequencing data, approximately equal to the total amount of DNA isolable from 10^9 *Escherichia* coli cells. In order to analyze these environmental samples, it is inevitable to perform amplification to fulfill the nucleic acid amount requirement by current pyrosequencing technologies. However, obtaining unbiased and evenly amplified whole metagenomic amplification products which are suitable for downstream sequencing is still difficult. Quality control is necessary before sending the products to final sequencing. Clone library construction combining randomly picked-clone sequencing was proved to be a cost-effective method to examine the quality of the amplification products and to ensure the efficiency of next-generation sequencing (Wu et al., 2011).

In the previous study, Wu et al., (Wu et al., 2011) reported a new protocol to isolate and amplify RNA from small amounts (nanogram level) of deep-sea microbial samples. The protocol included three key steps: total RNA isolation by modified method combining Trizol (Invitrogen, CA) and RNeasy methods, amplification with WT-Ovation™ Pico RNA Amplification System (NuGEN,

CA), and then conversion of single-strand cDNA to double-strand DNA with NuGEN WT-Ovation™ Exon Module. Based on the results from the clone library sequencing and the preliminary analysis of the Roche GS FLX Titanium chemistry pyrosequencing, it was concluded that the method is suitable for metatranscriptomic study of microbial samples of low cell density and high impurity. Moreover, in the whole-transcriptome amplification process, three different amplification methods Illustra GenomiPhi V2 DNA Amplification Method (GE Healthcare, NJ), QuantiTect® Whole Transcriptome method (QIAGEN, CA), and WT-Ovation™ Pico RNA Amplification System (NuGEN, CA) were evaluated. Compared with two other methods evaluated, the NuGEN method demonstrated better amplification and relatively higher reproducibility on whole-transcriptome amplification (Table 1). With more attention paid to community structure and functionality of microbial communities present in extreme environments, such as deep-sea, cold spring, and deep frozen earth where very low microbial cell density and high concentration of potential inhibitors are often expected, the protocol developed here could be very a useful tool in allowing direct next-generation based metatranscriptomic analysis. Related biotechnologies are still under development to enhance the whole metagenome amplification efficiency and to lower the starting mass of the next-generation sequencing platform to nanogram or even hundred picogram level.

The key to a comprehensive study of the transcriptome is the identification of the full set of transcripts (Martin & Wang 2011). Despite the increasing ease of obtaining enough amount of DNA sequences for high-throughput sequencing during metatranscriptomic analysis, it is still a great challenge to analyze the meta-data efficiently to obtain meaningful results, especially for those microorganisms without a high-quality reference genome (Singhal 2013). For the same reason, metatranscriptomic analysis could be biased and partial since the analysis can only access DNA sequences with gene annotation or prediction (Martin & Wang 2011). Currently, the sequence reads obtained from next-generation sequencing platforms (except Sanger sequencing), including 454 pyrosequencing, Illumina, SOLiD, and Ion Torrent, are typically short (50 – 500 bps) (Metzker 2010), hence it is necessary to assemble short reads when reconstructing the total transcriptome, a complicate and challenging step that requires computing systems with large memories to run parallel algorithms. Although several *de novo* transcriptome assembly methods for a single genome have been developed (Zhao et al., 2011; Surget-Groba & Montoya-Burgos 2010), efforts are still needed for analysis of meta-data of complex microbial communities.

Deep-Sea Microbial Community and Its Genetic Adaptation

The biomass in the photic zone is remarkably higher than that in the aphotic ocean since the photic surface ocean is the place where the majority of marine primary productions take place (Orcutt et al., 2011). On the other hand, the biomass concentration decreases by roughly 2 orders of magnitude in the aphotic ocean (Nagata et al., 2000). In the deep sea, microbial cell density can be as low as $10^3 - 10^4$ cells per ml of seawater (Kawagucci et al., 2012). Though cell abundances decrease with depth, the biodiversity of microbial communities in the deep-sea is surprisingly high. In a previous study, archaeal communities' diversity maintains a similar level or even higher level compared to that in the upper ocean level (Winter et al., 2009). Deep-sea microbial community studies indicate that the deep-sea lifestyle is composed of a series of very subtle adaptations to cold, high-pressure environment, instead of dramatic alterations of gene content or structure. Lauro et al., (2007) compared several deep-sea psychrophilic and barophilic bacterial isolates including *Moritella* sp. PE36 (Delong et al., 1997), *Psychromonas* sp. CNPT3 (Delong et al., 1997), *Photobacterium profundum* SS9 (Vezzi et al., 2005), *Carnobacterium* sp. AT7 (Lauro et al., 2007), and *Shewanella* sp. KT99 (Lauro et al., 2007) with their corresponding surface-water strains, and have found that the deep-sea psychrophilic and barophilic strains typically possess higher rRNA operon copies, genome size or above-average intergenic regions, all of which are suggestive of a rapid response to environmental changes (Klappenbach et al., 2000; Lauro & Bartlett 2008). Other deep-sea specific features include lower membrane saturation to maintain cell plasticity (Allen et al., 1999; Allen & Bartlett 2002), starvation endurance to fight against oligotrophic dwelling condition (Bidle & Bartlett 2001; Mahdhi et al., 2012), motility to quest for nutrients (Wirsen & Molyneaux 1999; Wang et al., 2008), and pressure-sensitive flagella assembly and motor system (Shibata et al., 2005; Qin et al., 2011).

Using high-throughput sequencing data, Sogin et al (2006) provided a comprehensive survey of the vertical zonation of phylogenetic distribution of the microbial community in the North Pacific Ocean and found that the percentage of *Gammaproteobacteria* increases with depth. Similar findings were also obtained in the Mediterranean Sea and the Greenland Sea (Zaballos et al., 2006). In another study, bacteria *Prochlorococcus*, *Verrucomicrobiales*, *Flexibacteraceae*, *Gammaproteobacteria*, *Alphaproteobacteria*, and *Delta-*

proteobacteria were found to be dominant in the ocean's surface water, while *Deferribacteres*; *Planctomycetaceae*, *Acidobacteriales*, *Gemmatamonadaceae*, *Nitrospina*, *Alteromonadaeceae* were enriched in deep-sea water (DeLong et al., 2006). In the study (Wu et al., 2013), it was found that chemolithoautotrophic and major nitrifier *Thaumarchaeota*, *Betaproteobacteria* and *Gammaproteobacteria*, which were recognized as ammonia oxidizing archaea (AOA) and ammonia oxidizing bacteria (AOB) that oxidize ammonia to nitrite (Pester et al., 2012), are enriched in the deep sea. Many of these AOA and AOB have a high ammonia affinity and can grow in extremely oligotrophic environments (Schleper & Nicol 2010). *Planctomycetes* are main members of the well-known anammox bacterial genus identified previously in the marine sub-oxic zone including *Rhodopirellula*, *Blastopirellula*, *Planctomyces*, *Pirellula*, *Candidatus Kuenenia*, *Gemmata* and *Isosphaera* (Kirkpatrick et al., 2006; Byrne et al., 2009). The presence of *Planctomycetes* is also an indicator that anaerobic ammonium oxidation (anammox), a globally important microbial process of the nitrogen cycle, may be another metabolic pathway supporting primary production in the low-oxygen, dark pelagic ocean. *Actinobacteria* and *Firmicutes*, two commonly seen deep-sea members, are believed to have adaptive advantage under low-nutrient conditions of the deep sea (Gärtner et al., 2011).

Except the phylogenetic features of deep-sea microbial community, microorganisms living in the deep-sea environments also possess special metabolic adaptation to the environments. In previous studies, bacteria living in the deep-sea have been shown to cope with the high hydrostatic pressure environment mainly through membrane phospholipid fatty acid unsaturation (Allen et al., 1999; Nagata et al., 2000; Allen & Bartlett 2002) and cell division (Bidle & Bartlett 2001; Ishii et al., 2005). Also, hydrostatic pressure was proven to be able to affect microorganisms' cellular processes including replication and translation (Welch et al., 1993; Ishii et al., 2005). By employing metagenomic and metatranscriptomic analysis, it has been found that in deep-sea environments, genes involved in functional categories including 'energy production and conversion', 'transcription', 'replication, recombination and repair', 'cell motility', 'inorganic ion transport and metabolism', 'secondary metabolism', 'signal transduction mechanism', and 'defense mechanism' are more enriched in deep-sea microbial communities (Konstantinidis et al., 2009), while genes involved in functional categories including 'cell division, chromosome partitioning', 'amino acid transport and metabolism', 'nucleotide transport and metabolism', 'carbohydrate transport and metabolism', 'coenzyme transport and metabolism', 'lipid transport and metabolism', 'translation and biogenesis', 'cell wall/membrane/envelope', 'protein turnover, chaperones', and 'intracellular trafficking and secretion' are less enriched in deep-sea microbial communities than in surface-ocean communities. The recent study (Wu et al., 2013) using integrated metagenomic and metatranscriptomic analysis of deep-sea microbial communities in the North Pacific Ocean reached similar conclusions except that in this study, genes related to 'cell wall/membrane/envelope' are more abundant in deep-sea than in surface water. The results were consistent with previous metagenomic studies that genes from abundant planktonic microbial populations are distinct between upper and lower levels of sea water (Frias-Lopez et al., 2008; Konstantinidis & DeLong 2008; Konstantinidis et al., 2009; Shi et al., 2011).

CONCLUSION

Nowadays, more taxonomic and functional studies of deep-sea microbial communities are conducted to unveil deep-sea microbial ecology. However, deep-sea environments are huge microbial reservoirs that still remain mostly unknown. The development and broad utilization of next-generation sequencing based metagenomic and metatranscriptomic analysis, and also the increasingly sophisticated computer modeling technologies make it possible to uncover the full taxonomical picture of microbial communities in special dwelling environments.

Recent studies found that prokaryotic microbes in deep-sea environments are well-adapted to the special dwelling environments after long evolution; carrying genetic features that enable them live and reproduce in the extreme environmental conditions. Recent progress in sequencing technologies is fueling a rapid increase in the number and scope of deep-sea microbial community-targeted studies. While metagenomic analysis can provide information on the taxonomic composition and metabolic potential of microbial communities in the deep sea, metatranscriptomics serves to unveil the actual metabolic activities of the communities at a specific time and location, and how those activities are changing in response to environmental and biotic parameters.

ACKNOWLEDGMENTS

The authors acknowledge the support of writing this book chapter and related research by the ASU NEPTUNE project to D. Meldrum.

REFERENCES

Abell GCJ, Robert SS, Frampton DMF, Volkman JK, Rizwi F, Csontos J, et al., (2012). High-Throughput Analysis of Ammonia Oxidiser Community Composition via a Novel, amoA-Based Functional Gene Array. *PLoS ONE* 7:e51542.

Allen E E, Facciotti D, Bartlett D H. (1999). Monounsaturated but not polyunsaturated fatty acids are required for growth of the deep-sea bacterium Photobacterium profundum SS9 at high pressure and low temperature. *Appl. Environ. Microbiol.* 65:1710–1720.

Allen Eric E, Bartlett Douglas H. (2002). Structure and regulation of the omega-3 polyunsaturated fatty acid synthase genes from the deep-sea bacterium Photobacterium profundum strain SS9. *Microbiology* (Reading, Engl.) 148:1903–1913.

Amann RI, Ludwig W, Schleifer KH. (1995). Phylogenetic identification and in situ detection of individual microbial cells without cultivation. *Microbiol. Rev.* 59:143–169.

Anantharaman K, Breier JA, Sheik CS, Dick GJ. (2013). Evidence for hydrogen oxidation and metabolic plasticity in widespread deep-sea sulfur-oxidizing bacteria. *Proc. Natl. Acad. Sci. U.S.A.* 110:330–335.

Aristegui J, GASOL JM, DUARTE CM, HERNDL GJ. (2009). Microbial oceanography of the dark ocean's pelagic realm. *Limnol. Oceanogr.* 54:1501–1529.

Bidle KA, Bartlett D H. (2001). RNA arbitrarily primed PCR survey of genes regulated by ToxR in the deep-sea bacterium Photobacterium profundum strain SS9. *J. Bacteriol.* 183:1688–1693.

Broemeling DJ, Pel J, Gunn DC, Mai L, Thompson JD, Poon H, et al., (2008). An instrument for automated purification of nucleic acids from contaminated forensic samples. *JALA Charlottesv* Va 13:40–48.

Byrne N, Strous M, Crépeau V, Kartal B, Birrien J-L, Schmid M, et al., (2009). Presence and activity of anaerobic ammonium-oxidizing bacteria at deep-sea hydrothermal vents. *The ISME Journal* 3:117–123.

Cassman N, Prieto-Davó A, Walsh K, Silva GGZ, Angly F, Akhter S, et al., (2012). Oxygen minimum zones harbour novel viral communities with low diversity. *Environ. Microbiol.* 14:3043–3065.

Delong E F, Franks DG, Yayanos A A. (1997). Evolutionary relationships of cultivated psychrophilic and barophilic deep-sea bacteria. *Appl. Environ. Microbiol.* 63:2105–2108.

DeLong Edward F, Preston CM, Mincer T, Rich V, Hallam SJ, Frigaard N-U, et al., (2006). Community genomics among stratified microbial assemblages in the ocean's interior. *Science* 311:496–503.

Demeke T, Jenkins GR. (2010). Influence of DNA extraction methods, PCR inhibitors and quantification methods on real-time PCR assay of biotechnology-derived traits. *Anal. Bioanal. Chem.* 396:1977–1990.

Deming JW, Reysenbach AL, Macko SA, Smith CR. (1997). Evidence for the microbial basis of a chemoautotrophic invertebrate community at a whale fall on the deep seafloor: bone-colonizing bacteria and invertebrate endosymbionts. *Microsc. Res. Tech.* 37:162–170.

Eloe EA, Fadrosh DW, Novotny M, Zeigler Allen L, Kim M, Lombardo M-J, et al., (2011). Going Deeper: Metagenome of a Hadopelagic Microbial Community. *PLoS ONE* 6:e20388.

Fang J, Shizuka A, Kato C, Schouten S. (2006). Microbial diversity of cold-seep sediments in Sagami Bay, Japan, as determined by 16S rRNA gene and lipid analyses. *FEMS Microbiol. Ecol.* 57:429–441.

Fenical W, Jensen PR. (2006). Developing a new resource for drug discovery: marine actinomycete bacteria. *Nature Chemical Biology* 2:666–673.

Frias-Lopez J, Shi Y, Tyson GW, Coleman ML, Schuster SC, Chisholm SW, et al., (2008). Microbial community gene expression in ocean surface waters. *Proc. Natl. Acad. Sci. U.S.A.* 105:3805–3810.

Fuhrman JA. (2002). Community structure and function in prokaryotic marine plankton. *Antonie Van Leeuwenhoek* 81:521–527.

Gärtner A, Blümel M, Wiese J, Imhoff JF. (2011). Isolation and characterisation of bacteria from the Eastern Mediterranean deep sea. *Antonie Van Leeuwenhoek* 100:421–435.

Gifford SM, Sharma S, Rinta-Kanto JM, Moran MA. (2011). Quantitative analysis of a deeply sequenced marine microbial metatranscriptome. *ISME J* 5:461–472.

Gilbert JA, Field D, Huang Y, Edwards Rob, Li W, Gilna P, et al., (2008). Detection of Large Numbers of Novel Sequences in the Metatranscriptomes of Complex Marine Microbial Communities. *PLoS ONE* 3:e3042.

Giovannoni SJ, Britschgi TB, Moyer CL, Field KG. (1990). Genetic diversity in Sargasso Sea bacterioplankton. , Published online: 03 May 1990; | doi:10.1038/345060a0 345:60–63.

Haefner B. (2003). Drugs from the deep: marine natural products as drug candidates. *Drug Discovery Today* 8:536–544.

Handelsman J. (2004). Metagenomics: application of genomics to uncultured microorganisms. *Microbiol. Mol. Biol. Rev.* 68:669–685.

Head IM, Saunders JR, Pickup RW. (1998). Microbial Evolution, Diversity, and Ecology: A Decade of Ribosomal RNA Analysis of Uncultivated Microorganisms. *Microb Ecol* 35:1–21.

Ishii A, Oshima T, Sato T, Nakasone K, Mori H, Kato C. (2005). Analysis of hydrostatic pressure effects on transcription in Escherichia coli by DNA microarray procedure. *Extremophiles* 9:65–73.

Jiménez DJ, Andreote FD, Chaves D, Montaña JS, Osorio-Forero C, Junca H, et al., (2012). Structural and functional insights from the metagenome of an acidic hot spring microbial planktonic community in the Colombian Andes. *PLoS ONE* 7:e52069.

Kawagucci S, Yoshida YT, Noguchi T, Honda MC, Uchida H, Ishibashi H, et al., (2012). Disturbance of deep-sea environments induced by the M9.0 Tohoku Earthquake. *Sci Rep* 2.

Kennedy J, Flemer B, Jackson SA, Lejon DPH, Morrissey JP, O'Gara F, et al., (2010). Marine metagenomics: new tools for the study and exploitation of marine microbial metabolism. *Mar. Drugs* 8:608–628.

Kirkpatrick J, Oakley B, Fuchsman C, Srinivasan S, Staley JT, Murray JW. (2006). Diversity and Distribution of Planctomycetes and Related Bacteria in the Suboxic Zone of the Black Sea. *Appl. Environ. Microbiol.* 72:3079–3083.

Klappenbach JA, Dunbar JM, Schmidt TM. (2000). rRNA operon copy number reflects ecological strategies of bacteria. *Appl. Environ. Microbiol.* 66:1328–1333.

Konstantinidis KT, Braff J, Karl DM, DeLong Edward F. (2009). Comparative metagenomic analysis of a microbial community residing at a depth of 4,000 meters at station ALOHA in the North Pacific subtropical gyre. *Appl. Environ. Microbiol.* 75:5345–5355.

Konstantinidis KT, DeLong Edward F. (2008). Genomic patterns of recombination, clonal divergence and environment in marine microbial populations. *ISME J.* 2:1052–1065.

Lashkari DA, DeRisi JL, McCusker JH, Namath AF, Gentile C, Hwang SY, et al., (1997). Yeast microarrays for genome wide parallel genetic and gene expression analysis. *PNAS* 94:13057–13062.

Lauro Federico M, Bartlett Douglas H. (2008). Prokaryotic lifestyles in deep sea habitats. *Extremophiles* 12:15–25.

Lauro Federico M., Chastain RA, Blankenship LE, Yayanos A. Aristides, Bartlett Douglas H. (2007). The Unique 16S rRNA Genes of Piezophiles Reflect both Phylogeny and Adaptation. *Appl. Environ. Microbiol.* 73:838–845.

Mahdhi A, Esteban MÁ, Hmila Z, Bekir K, Kamoun F, Bakhrouf A, et al., (2012). Survival and retention of the probiotic properties of Bacillus sp. strains under marine stress starvation conditions and their potential use as a probiotic in Artemia culture. *Res. Vet. Sci.* 93:1151–1159.

Martin JA, Wang Z. (2011). Next-generation transcriptome assembly. *Nat. Rev. Genet.* 12:671–682.

McGrath KC, Thomas-Hall SR, Cheng CT, Leo L, Alexa A, Schmidt S, et al., (2008). Isolation and analysis of mRNA from environmental microbial communities. *J. Microbiol. Methods* 75:172–176.

Metzker ML. (2010). Sequencing technologies — the next generation. *Nature Reviews Genetics* 11:31–46.

Muyzer G, Smalla K. (1998). Application of denaturing gradient gel electrophoresis (DGGE) and temperature gradient gel electrophoresis (TGGE) in microbial ecology. *Antonie Van Leeuwenhoek* 73:127–141.

Nagata T, Fukuda H, Fukuda R, Koike I. (2000). Bacterioplankton distribution and production in deep Pacific waters: Large-scale geographic variations and possible coupling with sinking particle fluxes. *Limnology and Oceanography* 45:426–435.

Ogura A, Lin M, Shigenobu Y, Fujiwara A, Ikeo K, Nagai S. (2011). Effective gene collection from the metatranscriptome of marine microorganisms. *BMC Genomics* 12:S15.

Orcutt B, Samarkin V, Boetius A, Joye S. (2008). On the relationship between methane production and oxidation by anaerobic methanotrophic communities from cold seeps of the Gulf of Mexico. *Environ. Microbiol.* 10:1108–1117.

Orcutt BN, Sylvan JB, Knab NJ, Edwards KJ. (2011). Microbial Ecology of the Dark Ocean above, at, and below the Seafloor. *Microbiol. Mol. Biol. Rev.* 75:361–422.

Orsi W, Edgcomb V, Jeon S, Leslin C, Bunge J, Taylor GT, et al., (2011). Protistan microbial observatory in the Cariaco Basin, Caribbean. II. Habitat specialization. *ISME J.* 5:1357–1373.

Pace N, stahl D, Lane D, Olsen G. (1986). The analysis of natural microbial populations by rRNA sequences. *Adv. Microb. Ecol.* 9:1–55.

Pace NR. (1997). A molecular view of microbial diversity and the biosphere. *Science* 276:734–740.

Parro V, Moreno-Paz M, González-Toril E. (2007). Analysis of environmental transcriptomes by DNA microarrays. *Environ. Microbiol.* 9:453–464.

Pearce DA, Newsham KK, Thorne MAS, Calvo-Bado L, Krsek M, Laskaris P, et al., (2012). Metagenomic analysis of a southern maritime antarctic soil. *Front Microbiol.* 3:403.

Pester M, Rattei T, Flechl S, Gröngröft A, Richter A, Overmann J, et al., (2012). amoA-based consensus phylogeny of ammonia-oxidizing archaea and deep sequencing of amoA genes from soils of four different geographic regions. *Environmental Microbiology* 14:525–539.

Poretsky RS, Bano N, Buchan A, LeCleir G, Kleikemper J, Pickering M, et al., (2005). Analysis of Microbial Gene Transcripts in Environmental Samples. *Appl. Environ. Microbiol.* 71:4121–4126.

Prieur D, Erauso G, Jeanthon C. (1995). Hyperthermophilic life at deep-sea hydrothermal vents. *Planet Space Sci* 43:115–122.

Qin Q-L, Li Y, Zhang Y-J, Zhou Z-M, Zhang W-X, Chen X-L, et al., (2011). Comparative genomics reveals a deep-sea sediment-adapted life style of Pseudoalteromonas sp. SM9913. *ISME J.* 5:274–284.

Rappé MS, Connon SA, Vergin KL, Giovannoni SJ. (2002). Cultivation of the ubiquitous SAR11 marine bacterioplankton clade. *Nature* 418:630–633.

Rusch DB, Halpern AL, Sutton G, Heidelberg KB, Williamson S, Yooseph S, et al., (2007). The Sorcerer II Global Ocean Sampling expedition: northwest Atlantic through eastern tropical Pacific. *PLoS Biol.* 5:e77.

Saidi-Mehrabad A, He Z, Tamas I, Sharp CE, Brady AL, Rochman FF, et al., (2012). Methanotrophic bacteria in oilsands tailings ponds of northern Alberta. *ISME J.*

Schena M, Shalon D, Davis R W, Brown P O. (1995). Quantitative monitoring of gene expression patterns with a complementary DNA microarray. *Science* 270:467–470.

Schleper C, Nicol GW. (2010). Ammonia-oxidising archaea--physiology, ecology and evolution. *Adv. Microb. Physiol.* 57:1–41.

Schmidt TM, DeLong E F, Pace NR. (1991). Analysis of a marine picoplankton community by 16S rRNA gene cloning and sequencing. . *Bacteriol.* 173:4371–4378.

Schrenk MO, Kelley DS, Delaney JR, Baross JA. (2003). Incidence and Diversity of Microorganisms Within the Walls of an Active Deep-Sea Sulfide Chimney. *Appl. Environ. Microbiol.* 69:3580–3592.

Shakya M, Quince C, Campbell JH, Yang ZK, Schadt CW, Podar M. (2013). Comparative metagenomic and rRNA microbial diversity characterization using archaeal and bacterial synthetic communities. *Environ. Microbiol.*

Shi Y, Tyson GW, DeLong Edward F. (2009). Metatranscriptomics reveals unique microbial small RNAs in the ocean's water column. *Nature* 459:266–269.

Shi Y, Tyson GW, Eppley JM, DeLong Edward F. (2011). Integrated metatranscriptomic and metagenomic analyses of stratified microbial assemblages in the open ocean. *ISME J.* 5:999–1013.

Shibata S, Alam M, Aizawa S-I. (2005). Flagellar filaments of the deep-sea bacteria Idiomarina loihiensis belong to a family different from those of Salmonella typhimurium. *J. Mol. Biol.* 352:510–516.

Singhal S. (2013). De novo transcriptomic analyses for non-model organisms: an evaluation of methods across a multi-species data set. *Mol. Ecol. Resour.*

Smith. (2003). Oceanography and Marine Biology: an annual review. CRC Press.

Stewart FJ, Ulloa O, Delong Edward F. (2011). Microbial metatranscriptomics in a permanent marine oxygen minimum zone. Environmental Microbiology. http://www.ncbi.nlm.nih.gov/pubmed/21210935 (Accessed December 1, 2011).

Surget-Groba Y, Montoya-Burgos JI. (2010). Optimization of de novo transcriptome assembly from next-generation sequencing data. *Genome Res.* 20:1432–1440.

Teske A, Durbin A, Ziervogel K, Cox C, Arnosti C. (2011). Microbial community composition and function in permanently cold seawater and sediments from an arctic fjord of svalbard. *Appl. Environ. Microbiol.* 77:2008–2018.

Tringe SG, Mering C von, Kobayashi A, Salamov AA, Chen K, Chang HW, et al., (2005). Comparative Metagenomics of Microbial Communities. *Science* 308:554–557.

Ulloa O, Canfield DE, DeLong Edward F., Letelier RM, Stewart FJ. (2012). Microbial oceanography of anoxic oxygen minimum zones. *PNAS* 109:15996–16003.

Urich T, Lanzén A, Qi J, Huson DH, Schleper C, Schuster SC. (2008). Simultaneous assessment of soil microbial community structure and function through analysis of the meta-transcriptome. *PLoS ONE* 3:e2527.

Valouev A, Ichikawa J, Tonthat T, Stuart J, Ranade S, Peckham H, et al., (2008). A high-resolution, nucleosome position map of C. elegans reveals a lack of universal sequence-dictated positioning. *Genome Res.* 18:1051–1063.

Venter JC, Remington K, Heidelberg JF, Halpern AL, Rusch D, Eisen JA, et al., (2004). Environmental Genome Shotgun Sequencing of the Sargasso Sea. *Science* 304:66–74.

Vezzi A, Campanaro S, D'Angelo M, Simonato F, Vitulo N, Lauro F M, et al., (2005). Life at depth: Photobacterium profundum genome sequence and expression analysis. *Science* 307:1459–1461.

Wang Fengping, Wang J, Jian H, Zhang B, Li S, Wang Feng, et al., (2008). Environmental adaptation: genomic analysis of the piezotolerant and psychrotolerant deep-sea iron reducing bacterium Shewanella piezotolerans WP3. *PLoS ONE 3*:e1937.

Welch TJ, Farewell A, Neidhardt FC, Bartlett D H. (1993). Stress response of Escherichia coli to elevated hydrostatic pressure. *J. Bacteriol.* 175:7170–7177.

Whiteley AS, Jenkins S, Waite I, Kresoje N, Payne H, Mullan B, et al., (2012). Microbial 16S rRNA Ion Tag and community metagenome sequencing using the Ion Torrent (PGM) Platform. *Journal of Microbiological Methods* 91:80–88.

Whitman WB, Coleman DC, Wiebe WJ. (1998). Prokaryotes: the unseen majority. *Proc. Natl. Acad. Sci. U.S.A.* 95:6578–6583.

Winter C, Kerros M-E, Weinbauer MG. (2009). Seasonal changes of bacterial and archaeal communities in the dark ocean : Evidence from the Mediterranean Sea. *Methods* 54:160–170.

Wirsen CO, Molyneaux SJ. (1999). A study of deep-sea natural microbial populations and barophilic pure cultures using a high-pressure chemostat. *Appl. Environ. Microbiol.* 65:5314–5321.

Wu J, Gao W, Johnson RH, Zhang W, Meldrum DR. Integrated metagenomic and metatranscriptomic analyses of microbial communities in the Meso- and Bathypelagic realm of North Pacific Ocean.

Wu J, Gao W, Zhang W, Meldrum DR. (2011). Optimization of whole-transcriptome amplification from low cell density deep-sea microbial samples for metatranscriptomic analysis. *J. Microbiol. Methods* 84:88–93.

Wyrtki K. (1962). The oxygen minima in relation to ocean circulation. *Deep Sea Research and Oceanographic Abstracts* 9:11–23.

Ye L, Zhang T, Wang T, Fang Z. (2012). Microbial structures, functions, and metabolic pathways in wastewater treatment bioreactors revealed using high-throughput sequencing. *Environ. Sci. Technol.* 46:13244–13252.

Youssef N, Steidley BL, Elshahed MS. (2012). Novel High-Rank Phylogenetic Lineages within a Sulfur Spring (Zodletone Spring, Oklahoma), Revealed Using a Combined Pyrosequencing-Sanger Approach. *Appl. Environ. Microbiol.* 78:2677–2688.

Zaballos M, López-López A, Ovreas L, Bartual SG, D'Auria G, Alba JC, et al., (2006). Comparison of prokaryotic diversity at offshore oceanic locations reveals a different microbiota in the Mediterranean Sea. *FEMS Microbiol. Ecol.* 56:389–405.

In: Deep Sea
Editor: Nigel T. Wilson

ISBN: 978-1-63321-637-2

Chapter 2

BIODIVERSITY AND LIFE-HISTORY STRATEGIES OF DEEP SEA MEGAFAUNA IN THE MEDITERRANEAN SEA

A. Quetglas*, B. Guijarro, J. L. López-Jurado, C. Orejas and E. Massutí
Instituto Español de Oceanografía, Centre
Oceanogràfic de les Balears, Palma, Spain

ABSTRACT

The Mediterranean Sea is a relatively small but deep basin of 1450 m of mean depth situated between 30° N and 46° N and 5.5° W and 36° E. The only natural connection with the Atlantic Ocean is the Strait of Gibraltar, a narrow and shallow channel of 14 km width and 320 m depth. Given that the high evaporation rates are not compensated by inputs from rainfall and river discharges, the Mediterranean acts as a concentration basin in which water circulation is dominated mainly by salinity and temperature differences (thermohaline forcing). The Mediterranean has been compared with a miniature ocean for studies on oceanographic and biological processes taking place at larger scales. Although representing only 0.8% of the area and less than 0.25% of the volume of the world's oceans, the Mediterranean includes about 6% of known world marine

* Corresponding author: A. Quetglas. Instituto Español de Oceanografía, Centre Oceanogràfic de les Balears, Moll de Ponent s/n, 07015 Palma, Spain. E-mail: toni.quetglas@ba.ieo.es.

fauna and 18% of world marine flora of which 28% are endemic. A total of 10000 to 12000 marine species (with 8500 species of macroscopic fauna) have been recorded. Despite its small dimensions, the Mediterranean coastal waters are a hot spot of biodiversity hosting a high percentage of endemic species. However, the diversity of deep megafauna is comparatively low, most probably due to the Gibraltar sill, which would act as a physical barrier for the entrance of larvae and deep-sea benthic organisms from the Atlantic. Although with some exceptions, there is a general decreasing pattern of biodiversity with increasing depth at different extent depending on the taxonomic group. Most megafaunal community studies have focused on fishes, decapod crustaceans and cephalopods and restricted to depths above 800 m. The epibenthic megafaunal communities below 200 m are still fairly unknown, yet cold-water corals communities have been described in recent years. Besides the environmental conditions typical of deep waters, Mediterranean deep-sea organisms must face high homeothermy (12.5-15.5°C), high salinity (38.0-39.5) and high oligotrophy (the eastern basin is one of the most oligotrophic areas of the world). The Mediterranean is among the most impacted seas in the world (e.g. crowding of coastal zones, biodiversity loss, overfishing and pollution). Climate change has been reported to affect deep-sea ecosystems and may have deleterious effects on the biodiversity of such sensitive habitats. In this chapter, we firstly introduce the Mediterranean Sea and describe its main oceanographic conditions. Secondly, we review current knowledge on the biodiversity and life-history strategies of Mediterranean deep-sea megafaunal groups such as fishes, crustaceans, cephalopods and cold-water corals. Finally, we discuss the main threats posed by human impacts on Mediterranean deep-sea ecosystems.

1. INTRODUCTION

The Mediterranean Sea constitutes the largest of the seas peripheral to the mean oceans (Tyler, 2003). It is landlocked between three continents (Europe, Asia and Africa), occupying an elongated basin of 4000 km length located between 30° N and 46° N and 5.5° W and 36° E . The only natural connection with the Atlantic Ocean is the Strait of Gibraltar, a narrow and shallow channel of 14 km width and 320 m depth.

It is also naturally connected with the Black Sea through the Dardanelles Strait, a narrow channel of a maximum depth of 103 m separating Europe from the mainland of Asia. Since 1869, it has also been artificially connected with the Red Sea by the Suez Canal. The average depth (1450 m) of the

Mediterranean is well shallower than the average depth of the world oceans (3850 m). Water circulation and sedimentary dynamics are highly complex owing to the landlocked position of the Mediterranean between continents, the restricted communication with the adjacent oceans and the numerous islands found throughout the two basins (Sarda et al., 2004; Danovaro et al., 2010).

The Mediterranean is divided into western and eastern basins, separated by the Strait of Sicily, a sill of about 400 m depth. These two main basins are further subdivided in different sub-basins. The western and eastern parts are subdivided into three (Alboran, Algero-Provençal and Thyrrenian basins) and two (Ionian and Levantine basins) submarine basins, respectively (Figure 1). The eastern basin reaches higher depths than the western basin, as reveals the maximum depths found in the Algero-Provençal (3420 m), Thyrrenian (3730 m), Ionian (5121 m) and Levantine (4384 m) basins (Vanney and Gennesseaux, 1985; Tyler, 2003).

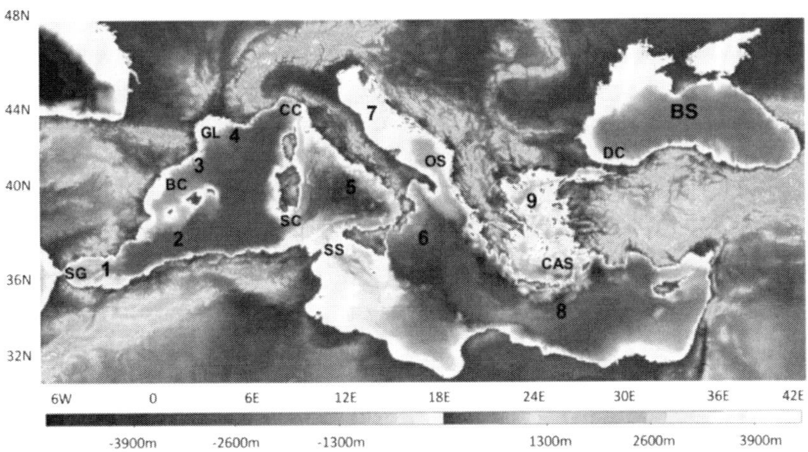

Source: International Ocean Commission, International Hydrographic Organization and British Oceanographic Data Centre –IOC/IHO/BODC-2003: Centenary Edition of the GEBCO Digital Atlas, published on CD-ROM. Liverpool, UK.

SG: Strait of Gibraltar, BC: Balearic Channels, CC: Corsica Channel, SC: Sardinia Channel, SS: Strait of Sicily, OS: Otranto Strait, CAS: Cretan Arc Straits, DC: Dardanelles Channel, GL: Gulf of Lions; BS: Black Sea.

Figure 1. Map showing the different sub-basins in both the western and eastern Mediterranean basins along with the main marine channels and straits. The western basin includes five subbasins: 1) Alboran Sea; 2) Algerian; 3) Balearic; 4) Liguro-Provençal; and 5) Tyrrhenian. The eastern basin contains four subbasins: 6) Ionian Sea; 7) Adriatic Sea; 8) Levantine; and 9) Aegean Sea.

Continental shelves account for less than 30% of the total Mediterranean area, whereas the bathyal grounds extends for about 60% of the whole basin and the abyssal plane constitutes the remaining 13%.

The Mediterranean waters are characterized by high salinity, ranging from 36.4 to 39.2 in the surface and exceeding 38.4 in depth. Temperature of superficial water varies between approximately 13° C in winter and about 26° C in summer (Figure 2).

Due to its geographical situation, continental climate and the weakness of winds during late spring and summer, the waters are strongly stratified, generating a significant mixed layer about 20 to 30 m depth. The thermocline situates below that layer, at about 30-150 m depth. In autumn and winter, however, the cold winds and low temperatures homogenize the water column and high oxygen concentrations can be found in the deep waters.

The Mediterranean has traditionally been considered an oligotrophic sea (Estrada, 1996) and the eastern basin is one of the most oligotrophic areas of the world (Psarra et al., 2000).

Owing to the high evaporation rates prevailing on Mediterranean latitudes, which are not compensated by inputs from rainfall and river discharges, the Mediterranean Sea acts as a concentration basin in which water circulation is dominated by salinity and temperature differences (thermohaline forcing). Water of relatively high salinity forms by evaporative concentration, sinks during winter convective processes to the intermediate and deep layers and outflows across the Gibraltar sill into the Atlantic.

The loss of deep Mediterranean water is over-compensated by the entrance of low salinity, superficial Atlantic waters through the Gibraltar Strait.

The basic geological configuration of the Mediterranean was shaped about 40 million years ago as a remnant of the Tethys Sea, an east-west trending ocean resulting from the collision between the African and Eurasian continents (Maldonado, 1985). During the late Miocene (Messinian), between six and five million years ago, the connection between the Atlantic and the Mediterranean through the Strait of Gibraltar was closed, giving rise to the "Messinian Salinity Crisis" (Maldonado, 1985).

Such closure prevented the inputs of Atlantic waters to offset the aforementioned high evaporation rates, transforming the Mediterranean into a series of large lakes in which precipitated extensive sequence of evaporites, most notably (1.0-3.5 km thick) in the deepest sea bottoms (Maldonado, 1985). According to McKenzie (1999), as much as 6% of the world's ocean salt was precipitated on the Mediterranean by evaporation.

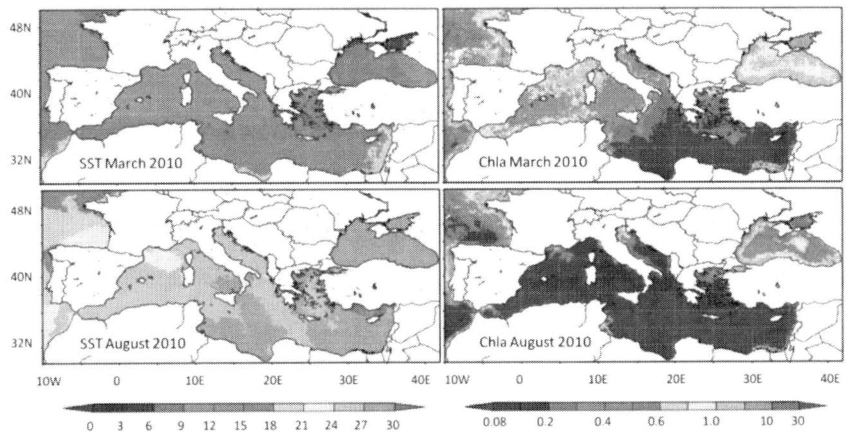

Source: NASA's Earth-Sun System Division, http://disc.sci.gsfc.nasa.gov/.

Figure 2. Satellite images of sea surface temperature (SST, °C) and chlorophyll a concentration (Chla, mg·m^{-3}) in the Mediterranean Sea during two contrasting oceanographic conditions in winter (March) and summer (August) of 2010.

This salinity crisis involved the complete disappearance of former deep benthic fauna which implies that the present deep Mediterranean organisms must originate from Atlantic migrations since 5.5 million years ago (Fredj and Laubier, 1985). No deep water species are considered Lessepsian, that is, has not entered the Mediterranean via the Suez Canal (Tyler, 2003).

Although representing only 0.8% of the surface and less than 0.25% of the volume of the world's oceans, the Mediterranean includes about 7% of known world marine fauna and 18% of world marine flora of which 28% are endemic (Fredj et al., 1992; CIHEAM, 2002). A total of 10000 to 12000 marine species (with 8500 species of macroscopic fauna) have been recorded. Up to 694 of these species are marine vertebrates, 540 are fishes and 363 of these species are exploited by fisheries. However, such estimates refer to the coastal waters whereas the knowledge of the life history and ecology of most deep-sea communities is currently rather scarce (Danovaro et al., 2010).

The Mediterranean has traditionally been compared with a miniature ocean for studies on oceanographic and biological processes taking place at larger scales and a useful framework to anticipate the response of the global oceans to different disturbances (Margalef, 1985; Lejeusne et al., 2010; Danovaro et al., 2010).

Tracking such responses is especially important, since the Mediterranean is one of the most impacted seas in the world (e.g. crowding of coastal zones, biodiversity loss, overfishing and pollution) and the climate change has been reported to affect even the deep-sea ecosystems and may have deleterious effects on the biodiversity of such sensitive habitats (Lejeusne et al., 2010).

2. Oceanography

2.1. Topographic Features and Meteorology

Much of the Mediterranean Sea is surrounded by important mountain systems and high plateaus. Such a complex topography is an effective barrier to dominant winds, which in turn affects the climate and the oceanic circulation of the Mediterranean. These mountain systems and valleys modify the atmospheric flow that would be expected from a geostrophic model, generating well known winds (e.g. mistral, tramontana, gregal, bora, meltemi) which converge in defined areas giving place to upwelling and convective processes due to localized cooling. Associated with these systems there are large rivers, such as the Ebro and Rhone in the western basin and the Po and Nile in the eastern basin. The Nile and Ebro deltas are landforms of great importance for their terrigenous contributions, biological production and shelf circulation effects. The Mediterranean is studded with islands that have an important role in the oceanic circulation: Cyprus, Crete, Rhodes and Malta in the eastern basin; Sardinia, Corsica, Sicily and the Balearic Islands in the western basin. As the insular shelf is narrow, the most important effect of the islands bathymetry is produced by the channels between islands and with the mainland. These channels play an important role in the regional circulation pattern and condition the exchange between sub-basins.

In a general sense, the isobath of 200 m is considered the boundary between the shelf and slope. The continental shelf at the western and eastern basins is relatively narrow, only the Tunisia Plateau close to the strait of Sicily and some areas like the Gulf of Lions, Northern Adriatic Sea, Aegean Sea, Nile and Ebro Deltas show a wider continental shelf. The continental slope is a narrow zone too, where depths higher than 2000 m are reached in a few miles. Along the Mediterranean coast, there are a variety of deep-sea canyons that generally have narrow, steep-sided, linear and sinuous channels, which allow continental sediments and shelf waters to be transported to oceanic basins.

These episodes occur when weather conditions are severe, leading to cascading processes that affect the intermediate and deep waters. These canyons can produce spatial variations in the shelf circulation off the coast (Masó et al., 1990).

At middle latitudes the general atmospheric circulation is zonal, in the direction of the geographic parallels with prevailing winds from west to east direction induced by the Azores high and the Iceland low pressure areas.

Three main features drive the meteorology of the Mediterranean: 1) the situation of the Azores high; 2) the cyclogenesis zone of the Alps; and 3) the fact that the Mediterranean is a relatively warm sea surrounded by highlands, makes it a source of water, heat and energy instability.

Changes in latitude of the Azores high give place to important weather differences in Europe and the Mediterranean area. The situation of this high may block the zonal flow (from west to east) and forward this to the north or south, resulting in three well defined weather regimes: 1) west-east flow or non blocking, when the Azores high is centered at 35º N over the Atlantic; 2) mixed flow, when the Azores high is shifted about 50º N; and 3) north-south flow, when the Azores high is located between 50º and 80º N, sending polar and Siberian winds to both the western and eastern basins.

Any of these regimes can change or enhance the marine circulation and the presence of north winds originates a very important process, the formation of deep Mediterranean water. In winter, the Gulf of Lions and the Ligurian Sea in the western basin and the Adriatic Sea in the eastern basin are frequently swept by these north winds, which are reinforced by the effect of the Alps, the main cause of cyclogenesis, generating the mistral-tramontana and bora wind systems. These winds have a dramatic effect on the vertical stability of the water column, which becomes denser at the surface and consequently begins to mix by convection; in some cases, this can affect the whole column from the surface to the bottom.

2.2. Hydrography, Water Masses and Circulation

In a wide sense, the marine circulation along the whole Mediterranean can be considered cyclonic, although factors such as the complex bottom topography, winds, submarine canyons and shelf-slope density fronts can complicate this simple circulation model in both basins (Figure 3).

All Mediterranean waters have an Atlantic origin. The different layers present along the water column are the same found in the Atlantic, just transformed into saltier and colder waters by several processes such as evaporation, mixing and convection and by the resident time spent in the Mediterranean. Depending on its position along the water column, such layers can be classified into surface, intermediate and deep waters.

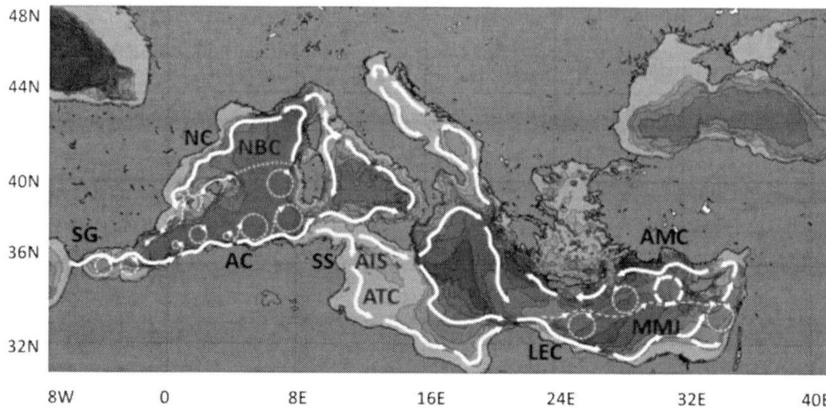

SG: Strait of Gibraltar; AC: Algerian Current; NC: Northern Current; NBF: Northern Balearic Front; SS: Strait of Sicily; AIS: Atlantic Ionian Stream; ATC: Atlantic Tunisia Current; LEC: Libyo Egyptian Current; AMC: Asia Minor Current; MMJ: Mid Mediterranean Jet.

Figure 3. Surface ocean currents in the Mediterranean Sea.

Throughout the year, the water column shows a marked thermal stratification during the summer period which is transformed during the autumn and become "quasi-homogenized" during the winter in large parts of the northern Mediterranean when temperatures remain approximately constant at 13°C. In spring, a considerable range of temperatures is observed at well separated areas due to differences in the increase of solar radiation among those areas. For this reason, a significant mixed layer can be gradually generated in summer reaching about 30 m depth. The main thermocline is clearly identified between 30 and 150 m depth.

At surface layers, the fresher Atlantic waters progressively lose its characteristics through mixing and evaporation during its travel to the east. For this reason, the salinity shows an eastward spatial gradient with values of 36.4 at Gibraltar and 39.2 in the Levantine basin.

These values are also associated with the resident times of the fresher and older Atlantic waters in the Mediterranean. On the other hand, at intermediate layers, this gradient is westward with the highest salinity values of about 39.2 at the Levantine basin and 38.4 in Gibraltar.

Deeper waters have temperature and salinity values of about 13.7°C and 38.7 in the Levantine basin, 12.85°C and 38.55 in the Adriatic South and 12.7°C and 38.4 in the western Mediterranean.

2.2.1. Surface Waters

The Atlantic water enters through the Gibraltar Strait into the Alboran Sea, where it is driven in a quasi permanent anticyclonic gyre in the west and a variable anticyclonic loop in the east of this sea and reaches Almeria-Oran front. This front stretches from the Spanish (Almeria) to the Algerian (Oran) coasts, separating the fresher Atlantic waters of the resident Mediterranean waters. This stream goes into the Algerian Basin and forms the Algerian Current, which circulates eastward along the Algerian slope to the Channel of Sardinia. This is a relatively narrow and deep current, with 30 to 50 km wide and 200-400 m depth at the coast (Benzohra and Millot, 1995), which can be destabilized by the coastal bathymetry generating coastal eddies.

Only the anticyclonic eddies can persist for a long time and, when detached from the current, they move freely and can reach about 100 km of diameter and 1000 m depth. Their eastward progression is topographically constrained by the bathymetry of the Sardinia Channel, where they are forced to move northward or anticlockwise into the Algerian basin. The boundary of the Atlantic waters into this basin is known as the North Balearic Front, which stretches between the Menorca and Corsica islands.

From here, a branch of Atlantic waters introduces into the Tyrrhenian Sea and another branch flows eastward through the Strait of Sicily and travels to the eastern Mediterranean. The northward currents along the west and east coasts of Corsica join at the Ligurian Sea and form the Northern Current, which flows southwestward along the continental slope of the Gulf of Lions and the Spanish coast to reach the Balearic channels. Then, this water crosses the channels and return to the Algerian basin (Pinot et al., 2002).

A portion of this current reinforces the Balearic Current that flows northeastward along the insular slope of these islands. At the northern coast of the western Mediterranean, this Atlantic water participates in the convective processes of winter, that lead to the formation of intermediate and deep waters.

On the other hand, the incoming Atlantic waters flowing into the eastern Mediterranean at the Sicily Strait, splits into two branches: the Atlantic Tunisian Current (ATC) that circulates around the Tunisian Coast and the Atlantic Ionian Stream (AIS) that flows eastward in the central and northern part of the strait.

In turn, the AIS bifurcate again with a principal vein flowing northeastward along the northern Ionian Sea and a weaker stream flowing along the Tunisian Shelf and introducing in the central Ionian Sea. These branches cross meandering the Ionian Sea and meet at ~20° E, giving place to the Libyo-Egyptian Current (LEC).

Thus, the Atlantic waters flow eastward along the continental slope, reaching the Middle East and the south of Cyprus coast and returns into the Asia Minor Current that flows along the southern coast of Turkey.

Within this cyclonic scheme there are quasi-permanent cyclonic eddies, such as the Rhodes and South Cyprus eddies, and some areas of anticyclonic circulation, such as the Mersa Matruh (over the Egypt coast) and Shikmona (over the Levantine coast) eddies and the northern area of the Ionian Sea (Gerin et al., 2009).

Robinson et al. (1991) suggested a scheme in which the main feature in the Levantine basin was a cross-basin current that they named the Mid Mediterranean Jet (MMJ). This MMJ is depicted crossing northeastward from ~24° E to southeast of Cyprus, and bifurcating to delineate circulation features such as "Mersa Matruh" and "Shikmona". They did not identify any surface current along Libya and Egypt slope. Recent studies consider MMJ a branch of the LEC, that flows northeastward between Mersa Matruh, Rhodes, Cyprus and Shikmona eddies around their northern edges (Hamad et al., 2006; Taupier-Letage, 2008). As a result of this circulation and the high evaporation rate, Atlantic waters are gradually transformed into saltier surface waters, reaching values of about 39.2 at the Levantine sub-basin.

2.2.2. Intermediate Waters

Most water masses are formed at the surface as a result of climatic and oceanographic processes such as evaporation and cooling, which generate increases in the water density. The highly dense surface waters sink down to their corresponding densities. The Levantine Intermediate Water (LIW) is the main intermediate water of the Mediterranean. It is formed in the eastern Mediterranean in several areas of the Levantine Basin and the southern Aegean Sea in winter under the influence of episodes of dry and cold north winds. Its core depth fluctuates from 50 to 600 m and its temperature and salinity range from 14.0 to 16.0°C and 38.70 to 39.20 respectively. The LIW have a relatively low oxygen content of less than 5 ml/l owing to its slow westward spreading and long renewal time (Manca et al., 2004). The LIW enters into the western Mediterranean through the Sicily Strait, circulating at intermediate depths where occupies a wide layer between 300 and 700 m and it is characterized by an absolute maximum of salinity, an absolute minimum of dissolved oxygen and a relative maximum of temperature in depth. Into this basin, the LIW describes a large cyclonic path along the northern coast, beginning into the Tyrrhenian Sea, which it leaves by the Sardinia Channel, border its south slope and turn sharply northward along the west coast of this island and Corsica to reach the Ligurian Sea and the Gulf of Lions. From this zone, it flows southward along the continental slope towards the Balearic Islands, crossing their channels and entering the Algerian sub-basin to finally reach the Alboran Sea (Katz, 1972). The LIW is a main component of the Mediterranean Outflow Water (MOW) into the Atlantic.

Another intermediate water mass is generally found at the western basin, the Western Intermediate Waters (WIW), that form seasonally in winter at the northern part of the western Mediterranean due to surface cooling and convection processes. Its main formation areas are the Gulf of Lions and the northern continental shelf of the Ligurian and Catalonian coasts. These waters are located between the surface waters and the LIW in a very varying thickness layer, which is characterized by an absolute temperature minimum (< 13ºC), but exhibit a wide range of salinity between 37.7 and 38.3 (Vargas et al., 2011). The WIW spreads southward, crossing the Balearic channels and entering into the Algerian sub-basin during spring and summer (Perkins and Pisteck, 1992); its presence is detected into the Alboran Sea during the late autumn and winter, where eventually disappears.

2.2.3. Deep Waters

Finally, the denser Deep Waters (DW) are found below the LIW and can be detected at depths from 800 m down to the bottom. These waters are formed in the winter processes of convection in different areas, such as the north of the Adriatic Sea and the Aegean Sea in the eastern basin and the Gulf of Lions and Ligurian Sea in the western basin. Its distribution from the northern regions where it forms is however poorly known, but the DW occupies the whole bottom layer of the Mediterranean. Deep water formation requires three phases: preconditioning, violent mixing, and sinking and spreading (MEDOC Group, 1970).

The preconditioning is favored by the atmospheric forcing and the regional cyclonic circulation, thus winds from north component displace the surface waters and replace them by colder and saltier sub-surface waters. The violent mixing needs cold and intense north winds acting over the new surface waters which generate heat and water vapor loss to the atmosphere; as a result of cooling, these waters become denser and mix with the saltier and warmer sub-surface waters. This deepening can affect to the entire water column and homogenize it. The spreading of these DW is not well known, but may follow the movement of other water masses along the slopes, occupying the entire seabed of both basins. Deep waters that reach the Alboran Sea are part of the Mediterranean waters that overflow to the Atlantic Ocean (MOW).

Although deep water formation occurs mostly in the open sea, strong wind events flowing over the continental shelf can lead to the formation of deep waters. When this occurs, the new deep waters slide down the shelf and the slope as a cascade reaching the seabed. This cascading is favored and more intense at the canyon areas and it is very important from the point of view of biogeochemical contributions to deep waters.

The main formation area of DW at the eastern basin is the north Adriatic Sea and these waters are known as the Adriatic Deep Waters (ADW); secondary sources are some areas of the Aegean Sea. From these areas, DW is distributed along the eastern basin occupying the entire bottom depths and forming the Eastern Mediterranean Deep Waters (EMDW). In the western basin, the main formation areas are the Gulf of Lions and the nearby Ligurian Sea, which give rise to the Western Mediterranean Deep Waters (WMDW); a secondary origin is the Tyrrhenian Sea, generating the Tyrrhenian Deep Waters (TDW). These later waters are lightly saltier and warmer than the former by the important contribution of the LIW. The distribution of the WMDW is poorly known, although they occupy the whole bottoms of the Algerian Basin and Alboran Sea. The sill in the Strait of Sicily prevents a direct communication between the EMDW and WMDW but coupling is achieved via the LIW layer (Lascaratos et al., 1999). Along the 1990s there was a drastic change in the EMDW, which shifted the main formation area of DW to the South Aegean Sea, and was known as the Eastern Mediterranean Transient (EMT). An abrupt change in the structure of the WMDW has also been observed after the winter of 2005 (López-Jurado et al., 2005), dubbed as the Western Mediterranean Transition (WMT), that spreads rapidly to the whole basin from its creation area (Figure 4).

During the last decade, the severe winter of 2004-2005 caused a high heat-loss at the area of WMDW formation, triggering the production of immense amounts of cold dense waters that quickly spread along the whole western Mediterranean basin creating a complex structure that is easy identifiable on the T/S diagram (López-Jurado et al., 2005). Three source waters contribute to this structure (Figure 4): 1) cascading, the coldest WMDW mode; 2) new, open ocean convention, the saltier/warmer WMDW mode; and 3) older WMDW mode, between the other two.

Oceanographic data collected during the 2005-2009 period in the Balearic Sea revealed that this termohaline anomaly persisted and was extended to the whole western Mediterranean basin (Schröder et al., 2010). It can be a good example of the kind of ocean circulation modification that has been induced by climate change scenarios (CIESM, 2009).

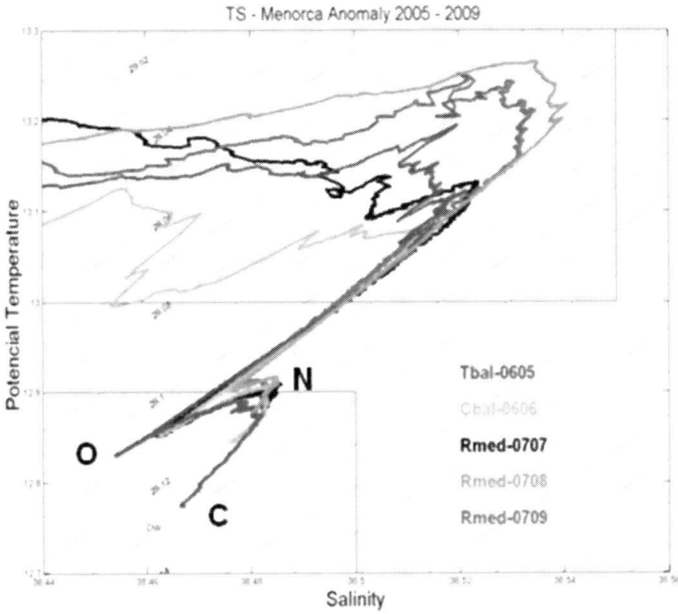

Names on the figure stand for the different scientific surveys developed between June 2005 and July 2009.

Figure 4. The Western Mediterranean Transition (WMT) as detected for the first time and its annual evolution from 2005 to 2009 (López-Jurado et al., 2005). Three different source waters are identified: 1) cascading, the coldest mode (C); 2) new, open ocean convention, the saltier/warmer mode (N); and 3) older mode, between the other two (O).

A first signature of the huge WMT seems to be evident in the Mediterranean outflow water (MOW) (Garcia Lafuente et al., 2007). In some aspects, the two events (EMT and WMT) are similar (abrupt warming and salting of the deep layers in the whole basin, uplifting by several hundreds of meters the old deep water), but there are also a number of crucial differences; there is no identified additional deep water source in the western Mediterranean and no circulation changes have been reported so far (CIESM, 2009).

3. Marine Biodiversity

Despite the Mediterranean coastal waters are a hot spot of biodiversity hosting a high percentage of endemic species, the diversity of deep megafauna is comparatively low (Danovaro et al., 2010). The deep Mediterranean megafauna is significantly impoverished when compared with similar Atlantic and Pacific ecosystems (Massuti et al., 2004). Most studies argue that such impoverishment is primarily due to the shallow Gibraltar sill, which would act as a physical barrier for the entrance of deep-sea organisms from the richer Atlantic (Fredj and Laubier, 1985; Bouchet and Taviani, 1992; Sarda et al., 2004; Danovaro et al., 2010). However, as the Gibraltar sill is not an impenetrable barrier for some species, other environmental factors might be involved, most notably the relatively high temperatures and salinity and the scarcity of food supply at deep waters. Indeed, Bouchet and Taviani (1992) suggested that the high temperature and salinity of the Mediterranean would prevent the metamorphosis or reproduction of those larval stages that succeeded in crossing the Gibraltar sill. As a result, these authors hypothesized that the Mediterranean deep-sea fauna consists of reproductively sterile pseudopopulations that are constantly replenished through the larval inflow from Atlantic mother populations. Nevertheless, the findings of gravid benthic crustaceans and fishes at great depths in the Levantine Sea (Goren and Galil, 1997; Fishelson and Galil, 2001) do not agree with this hypothesis.

In an excellent recent review, Danovaro et al. (2010) summarized all available information on benthic biodiversity (Prokaryotes, Foraminifera, Meiofauna, Macrofauna and Megafauna) from Mediterranean deep-sea ecosystems and drew two important conclusions. First, all of the biodiversity components, except Bacteria and Archaea, displayed a decreasing pattern with increasing depth, but to a different extent for each component. None of the groups analyzed by those authors displayed the unimodal pattern of diversity increasing with depth and peaks at mid-slope (Rex, 1981; Levin et al., 2001; Rex et al., 2005). Second, in contrast to what was expected from the sharp decrease in organic carbon fluxes and reduced faunal abundance, the deep-sea biodiversity of both the eastern and western basins was similar. This indicates that though important, the effects of food supply on deep biodiversity are compensated or masked by other factors, which can give rise to different responses depending on the taxonomic group and thus causing unpredictable biotic responses. Furthermore, despite megafauna composition displayed differences between the western and eastern basins, similar bathymetric patterns of species richness have been observed for fishes and crustaceans (D'Onghia et al., 2004; Company et al., 2004).

Danovaro et al. (2010) also furnished preliminary estimates of the number of species in the deep Mediterranean Sea considering different bathymetric strata. For the 200-1000 m depth stratum, about 650 species of the Eukaria domain were found and megafauna and nematodes contributed almost equally to total diversity, whereas the contribution of foraminifers and macrofauna was lower. On the 1000-2000 m depth stratum, the number of species reduced by almost a half with a relative increase of meio- and macrofauna groups. On deeper bottoms, diversity was further reduced by about 40% with a significant relative increase of foraminifers and meiofauna. The authors eventually concluded that approximately 66% of the total deep-sea Mediterranean diversity remains undiscovered.

Most megafaunal community studies in the Mediterranean have focused on fishes, decapod crustaceans and cephalopods and restricted to depths above 800 m. The epibenthic megafaunal communities below 200 m are still fairly unknown, yet important efforts have been made in the last decade and recently emblematic deep benthic communities, as the ones dominated by cold-water corals, have been described in several Mediterranean locations. According to Sarda et al. (2004), the vast majority of the deep Mediterranean basin is virtually unknown and the eastern basin remained one of the most poorly studied areas of the world until the early 1990s. Food supplies to the seafloor are the most important factors driving the abundance and diversity of deep-sea macro- and megafauna (Gage, 2003).

Besides the rough environmental conditions characteristics of deep waters, Mediterranean deep-sea organisms must face high homeothermy, high salinity and high oligotrophy. To cope with such conditions, megafaunal organisms have adopted different morphological and/or life-history adaptations. In the following sections, the available information on diversity patterns and ecological strategies of different megafaunal components (fishes, crustaceans, cephalopods and corals) from the Mediterranean deep-sea waters are reviewed.

3.1. Fishes

Fishes, jointly with decapods crustaceans, are the most important organisms in the nekton-benthic communities of the deep Mediterranean, both in terms of number of species as well as in terms of abundance and biomass (Figure 5). During the last decades, the deep sea ichthyofauna of the Mediterranean has been extensively studied, mainly based on bottom trawl sampling. Specially in the western and central basins, for both community structure (e.g. Stefanescu et al., 1993; Moranta et al., 1998; D'Onghia et al., 1998a) and distribution and biology of the main species are well know up to 2000 m depth (e.g. Massutí et al., 1995, 1996; Morales-Nin et al., 1996a,

1996b; D'Onghia et al., 1996, 2000, 1998b; Carrasón and Matallanas, 1998, 2001, 2002; Rotllant et al., 2002, Fernandez-Arcaya et al., 2012). Deep-sea fishes of the eastern Mediterranean have poorly been studied, and limited to 1000 m depth in the most-western part of this basin (Kallianiotis et al., 2000; Madurell et al., 2004; Madurell and Cartes, 2006), although some records of deep sea fishes captured with beam trawl up to 1500 m depth have been made off the most-eastern coast of the Mediterranean (Galil and Goren, 1994).

However, studies conducted during the last decade over a large spatial scale in the Mediterranean deep-sea, has allowed to improve this knowledge, not only in the eastern basin, but also in the western and central Mediterranean, by increasing the bathymetric range surveyed up to the deepest parts of these basins (D'Onghia et al., 2004a,b,c; Tecchio et al., 2011a,b). From more than 600 marine fishes catalogued in the Mediterranean (Quignard and Tomasini, 2000), only around 100 can be considered as deep-sea fishes, around 70 of them are nekto-benthic species (Table 1) and around 35 are meso- and bathy-pelagic species (Table 2).

Compared to the Atlantic, diversity and density of the deep water fish fauna of the Mediterranean is low. The factors that cause these differences are likely to be complex, including historical factors, energetic constraints and physical conditions (Tortonese, 1960; Fredj and Maurin, 1987; Bouchet and Taviani, 1992; Quignard and Tomasini, 2000).

Surface productivity and its arrival to the bottom, which has been estimated as an order of magnitude higher in the North Atlantic than in the Mediterranean (Berger et al., 1989; Miquel et al., 1994), as well as the more constant physical environment of the deep Mediterranean and the recent origin of its fauna (Pérès, 1985), have also been argued as important factors. The marine fauna of the Mediterranean is relatively young with respect to the fauna of the outer Atlantic (Fredj and Laubier 1985; Tyler 2003).

During the Messinian event, species from the northern Atlantic entered the Mediterranean, re-colonizing it. Thus, the species that now inhabit the continental slopes and abyssal basins of the Mediterranean are originally species that lived above 320 m depth (the maximum depth of the Gibraltar sill) and re-colonized the Mediterranean.

This is reflected in the general eurybathicity and widespread distribution of the majority of fishes living below a depth of 1000 m (Moranta et al., 1998; D'Onghia et al., 2004a) and for some crustacean species as well (Company et al., 2004).

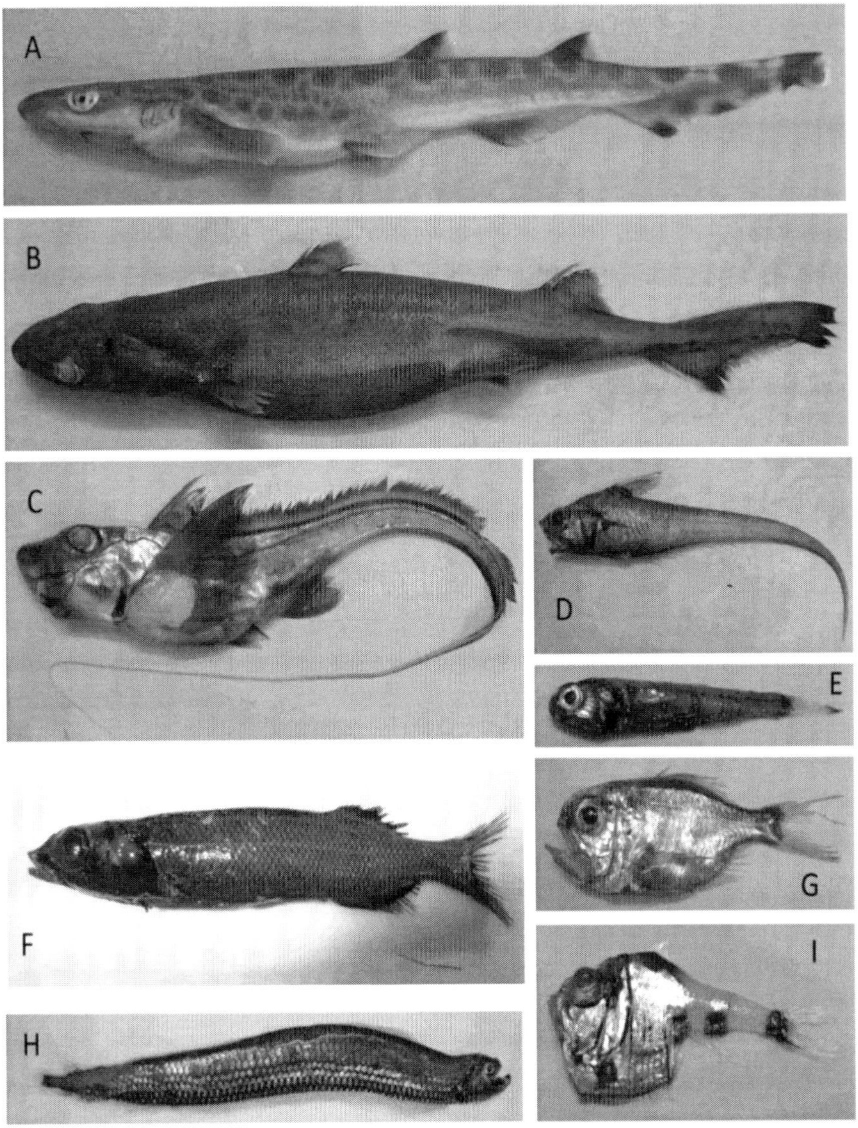

Figure 5. Pictures of some deep-sea fish species from the Mediterranean Sea: the elasmobranchs *Galeus melastomus* (A), *Etmopterus spinax* (B) and *Chimaera monstrosa* (C); and the teleosts *Nezumia aequalis* (D), *Ceratoscopelus maderensis* (E), *Alepocephalus rostratus* (F), *Hoplostethus meditarraneus* (G), *Stomias boa* (H) and *Argyropelecus hemygimnus* (I).

Table 1. Nekto-benthic fish species collected in the western (W; 400-3000 m depth), central (C; 600-4000 m depth) and eastern (E; 600-3000 m depth) deep-sea Mediterranean

Family	Species	W	C	E
Hexanchidae	*Hexanchus griseus*	+	+	+
Scyliorhinidae	*Galeus melastomus*	+	+	+
	Scyliorhinus canicula	+		
Oxynotidae	*Oxynotus centrina*		+	
Squalidae	*Centrophorus uyato*	+		
	Centrophorus granulosus			+
	Dalatias licha	+	+	+
	Etmopterus spinax	+	+	+
	Squalus blanvillei	+		
Somniosidae	*Centroscymnus coelolepis*	+		
	Somniosus rostratus			+
Rajidae	*Raja naevus*	+		
	Raja oxyrinchus	+		
	Raja clavata	+		
Chimaeridae	*Chimaera monstrosa*	+	+	+
Alepocephalidae	*Alepocephalus rostratus*	+		
Chlorophthalmidae	*Bathypterois mediterraneus*	+	+	+
	Chlorophthalmus agassizii	+		+
Nettastomatidae	*Nettastoma melanurum*	+	+	+
Congridae	*Conger conger*	+		+
Synaphobranchidae	*Dysomma brevirostre*	+		
Halosauridae	*Halosaurus ovenii*	+		
Notacanthidae	*Notacanthus bonapartei*	+	+	+
	Polyacanthonotus rissoanus	+		+
Macrouridae	*Coelorinchus caelorhincus*	+		
	Coelorinchus mediterraneus	+	+	+
	Chalinura mediterranea	+	+	+
	Coryphaenoides guentheri	+	+	+
	Hymenocephalus italicus	+	+	+
	Nezumia aequalis	+		
	Nezumia sclerorhynchus	+	+	+
	Trachyrincus scabrus	+	+	

Family	Species	W	C	E
Gadidae	*Gadiculus argenteus*	+		
	Antonogadus megalokynodon	+		
	Merluccius merluccius	+		+
	Micromesistius poutassou	+	+	+
	Molva dypterygia macrophthalma	+		
	Phycis blennoides	+	+	+
	Phycis phycis	+		
Moridae	*Eretmophorus kleinenbergi*	+		
	Lepidion guentheri	+		
	Lepidion lepidion	+	+	+
	Mora moro	+	+	+
	Rhynchogadus hepaticus	+		
Trachichthyidae	*Hoplostethus mediterraneus*	+	+	+
Apogonidae	*Epigonus constanciae*			+
Apogonidae	*Epigonus denticulatus*	+		
	Epigonus telescopus	+	+	
Sparidae	*Pagellus bogaraveo*	+	+	
Trichiuridae	*Lepidopus caudatus*	+		
Callionymidae	*Synchiropus phaeton*	+		
Zoarcidae	*Melanostigma atlanticum*	+	+	
Bythitidae	*Cataetyx alleni*	+		
	Cataetyx laticeps	+	+	+
Ophidiidae	*Benthocometes robustus*	+		
Carapidae	*Echiodon dentatus*	+		
Zoarcidae	*Melanostigma atlanticum*	+		
Bythitidae	*Cataetyx alleni*	+		
	Cataetyx laticeps	+	+	+
Ophidiidae	*Benthocometes robustus*	+		
Carapidae	*Echiodon dentatus*	+		
Scorpaenidae	*Helicolenus dactylopterus*	+		+
	Scorpaena elongata	+		
	Trachyscorpia cristulata echinata	+		
Triglidae	*Trigla lyra*	+		
Peristediidae	*Peristedion cataphractum*	+		+

Table 1. (Continued)

Family	Species	W	C	E
Liparididae	*Paraliparis leptochirus*	+		
Scophthalmidae	*Lepidorhombus boscii*	+		+
	Lepidorhombus whiffiagonis	+		+
Cynoglossidae	*Symphurus ligulatus*	+	+	+
	Symphurus nigrescens	+	+	
Lophiidae	*Lophius piscatorius*	+		
	Lophius budegassa	+		

From Moranta et al. (1998), Massutí and Moranta (2003), Massutí et al. (2004), D'Onghia et al. (2004) and Tecchio (2013).

Table 2. Meso-pelagic fish species collected with pelagic and bottom trawl in the deep-sea Mediterranean

Family	Species
Gonostomatidae	*Cyclothone braueri*
	Cyclothone pygmaea
Sternoptychidae	*Argyropelecus hemigymnus*
Sternoptychidae	*Maurolicus muelleri*
Phosichthyidae	*Ichthyococcus ovatus*
	Vinciguerria attenuata
Stomiidae	*Bathophilus nigerrimus*
	Chauliodus sloani
	Stomias boa
Argentinidae	*Argentina shyraena*
	Glossanodon leioglossus
Evermannellidae	*Evermannella balbo*
Paralepididae	*Lestidiops jayakari*
	Notolepis risso
	Paralepis speciosa
Myctophidae	*Benthosema glaciale*
	Electrona risso
	Hygophum benoiti
	Hygophum hygomii
	Myctophum punctatum
	Symbolophorus veranyi

Family	Species
	Ceratoscopelus maderensis
	Diaphus holti
	Diaphus metopoclampus
	Diaphus rafinesquei
	Lampanyctus crocodilus
	Lampanyctus pusillus
	Lobianchia dofleini
	Notoscopelus bolini
	Notoscopelus elongatus
Nemichthydae	*Nemichthys scolopaceus*
Macroramphosidae	*Macroramphosus scolopax*
Regalecidae	*Regalecus glesne*
Caproidea	*Capros aper*
Centrolophidae	*Centrolophus niger*
	Schedophilus medusophagus
Liparidae	*Paraliparis leptochirus*

From Moranta et al. (1998), Massutí et al. (2004), D'Onghia et al. (2004), and Tecchio (2013).

There are other differences between the Mediterranean and the Atlantic deep-sea fish assemblages. Where the same species occur, those in the Mediterranean tend to attain smaller adult sizes (Figure 6). The small size of Mediterranean deep-sea fishes has been discussed by Tortonese (1960) and Stefanescu et al. (1992).

According to these authors, this phenomenon is very complicated and should be a result of a combination of different factors, such as latitudinal gradients on fish size and ecological factors affecting body size, among others.

More limited resources and a high temperature in the Mediterranean (~13°C compared to 10 to 4°C in the North Atlantic) could also be assumed to play an important role. The dominant families in the assemblages are also different (Massutí et al., 2004). While different fish families are found at different depth stata in the Mediterranean (Gadidae on the upper slope, Moridae on the middle slope and Alepocephalidae in the lower slope), the family Macrouridae dominates along the bathymetric gradient in the North Atlantic (Figure 7).

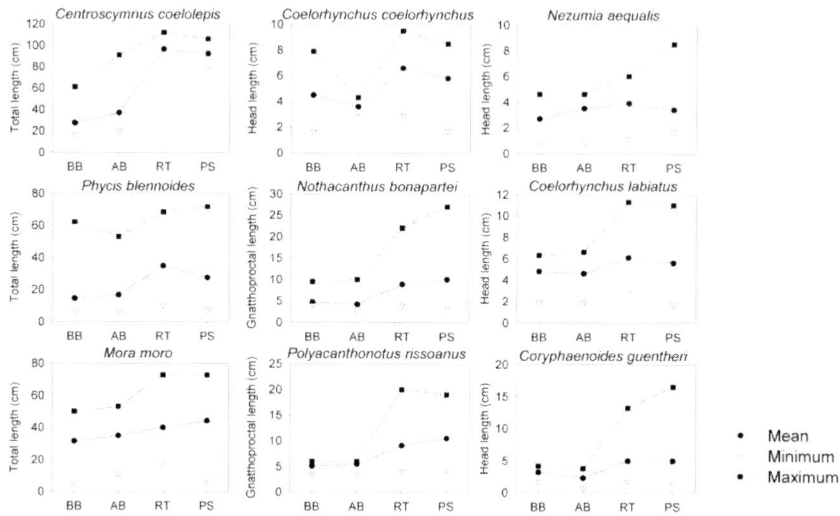

Adapted from Massuti et al. (2004).

Figure 6. Average (arithmetic mean), minimum and maximum length of some deep sea fishes in the western Mediterranean (BB: Balearic sub-basin off north-eastern Iberian Peninsula; AB: Algerian sub-basin southern Balearic Islands) and North Atlantic (RT: Rockall Trough; PS: Porcupine Seabight).

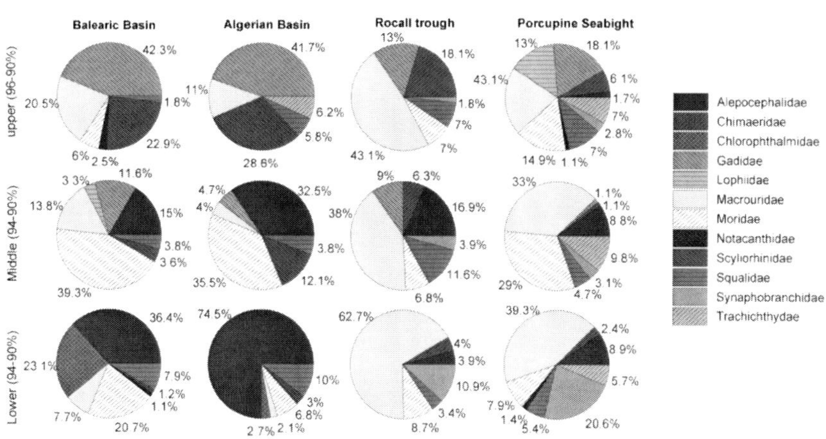

Adapted from Massuti et al. (2004).

Figure 7. Biomass composition (as percentage) for the most important families of deep sea fish assemblages in the western Mediterranean (BB: Balearic sub-basin off north-eastern Iberian Peninsula; AB: Algerian sub-basin southern Balearic Islands) and North Atlantic (RT: Rockall Trough; PS: Porcupine Seabight).

Differences in food availability and the partitioning of the main trophic resources within these ecosystems may also explain some of these differences.

Within the context of the oligotrophic Mediterranean conditions, a decreasing trend in the diversity, density and mean size of deep-sea fishes along the west-to-east axis has been observed, which could be linked to the fact that the western basin shows a higher productivity than the eastern one (Danovaro et al., 1999; Maynou and Cartes, 2000).

Geographic variations in the species dominance have also been observed, probably related to the differences in the trophic webs. Moranta et al. (1998) have reported that *Alepocephalus rostratus*, a species which feed mainly on macro-plankton (Carrasón and Matallanas, 1998), is more abundant south of the Balearic Islands than off north-eastern Iberian Peninsula, where the species *Lepidion lepidion* and *Bathypterois mediterraneus* which mainly feed on benthic decapods (Carrasón et al., 1997) and bentho-pelagic crustaceans (Carrasón and Matallanas, 2001), respectively, are more abundant. While in the Balearic Islands the food of planktonic origin is more important, the big submarine canyons along the north-eastern Iberian Peninsula increase the role of benthic and bentho-pelagic preys in the food webs (Cartes et al., 1994).

In general, a similar decreasing trend has also been reported along the bathymetric gradient, in accordance with the general global pattern described by Rex et al. (2006). It is well known that in the deep eastern Mediterranean macro-faunal organisms are scarce, with abundance, biomass and diversity displaying an exponential decrease with depth, due to the extremely unfavourable trophic conditions (Tecchio, 2013).

An important feature of the deep-sea fish assemblages is the well known peak of biomass at 1000-1500 m depth (e.g. Moranta et al., 1998; D'Onghia et al., 2004; Tecchio et al., 2013), where some medium and large-sized species reach their greatest abundance (e.g. *Alepocephalus rostratus* and *Mora mora*) and some other species reach their greatest biomass (e.g. *Trachyrincus scabrus* and *Phycis blennoides*). According to Hecker (1990) and Stefanescu et al. (1993), this bathymetric zone can be considered a "transitional" zone, where an important species turnover occurs. This ichthyofauna zonation has been attributed to the trophic resources available above and below 1200 m depth, linked to the distribution of meso-pelagic organisms (e.g. Stefanescu et al., 1993). The larger species with high energy requirements, such as *A. rostratus, T. scabrus, M. moro* and *Galeus melastomus*, are replaced at greater depths by smaller species which exploit a broad spectrum of bentho-pelagic preys, such as *Bathypterois mediterraneus, Chalinura mediterranea* and *Coryphaenoides guentheri*.

The impact of the deep water bottom trawl fishery targeting decapod crustaceans down to 800-900 m depth, may also explain this biomass pattern, since the peak observed around 1200 m depth could be due to the biomass removal at shallower grounds. In fact, the main differences in biomass spectra between Mediterranean and Atlantic deep sea fish assemblages have been detected on the upper slope; whereas small-size classes predominated in the Mediterranean, they were scarce in the Atlantic where these grounds are not so intensively exploited (Figure 8).

The predatory strategies and morphologies of deep-sea fishes minimize its foraging energetic cost in an environment limited in energy resources (Crabtree et al., 1991). Thus, it might be expected that nekto-benthic species have low rates of food consumption and utilize “sit-and-wait” or “ambush” feeding strategies, among others. Haedrich and Merrett (1992) showed that most nekton-benthic fishes are wide foragers or scavengers, which exploit allochthonous food sources from shallower layers, migrating mid-water fauna and small particles within the benthic boundary layer.

It has been found that meso-pelagic fauna is an important food source for some of these species, which preyed them close to the bottom or actively migrate up into the meso-pelagic zone to feed (e.g. Clarke and Merrett, 1972; Haedrich and Henderson, 1974; Bulman and Koslow, 1992).

Deep-sea fishes also show a variety of reproductive patterns, with both gonochoric (e.g. Liparidae and Zoarcidae) and hermaphroditic species (Chlorophthalmidae). There are species with low fecundity and several generations of oocytes in the ovary, with a less seasonal reproductive pattern, more adapted to oligotrophic conditions of the southern latitudes, and also species with high fecundity, eggs of same stage in the ovary and a seasonal spawning, more adapted to eutrophic seasonal conditions of the northern latitudes (Gage and Tyler, 1991). Similarly, there are species with numerous small eggs, which develop pelagically through larval stages (e.g. Synaphobranchidae, Notacanthidae and Gadiformes), but Alepocephalidae have generally much larger eggs, which imply demersal development (Merrett and Haedrich, 1997). Data on early life history are sparse or absent.

The current knowledge of the biology of deep-water species in the Mediterranean is still fragmentary, especially in aspects related to age composition, growth, reproductive characteristics and fecundity. However, we can conclude that species with different life histories are segregated by depth. Large fish with high energy requirements are found mainly on the upper and middle slope, while the more sedentary and/or smaller species with low energy costs seem to be better adapted to the poorer environment of the lower slope.

In the context of r-k ecological theory, the deep-sea fishes are typical k-strategists owing to several common traits, such as slow growth, long life span and delayed maturity. These bio/ecological traits should be considered in establishing management policies for the conservation of a fragile ecosystem as the Mediterranean deep-sea.

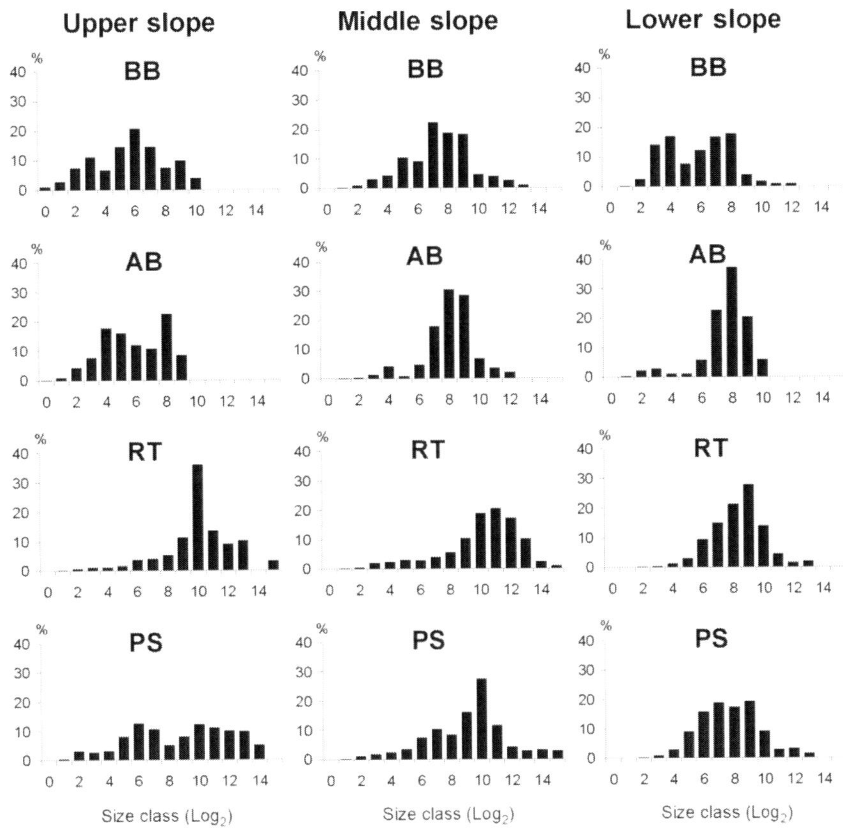

From Massuti et al. (2004).

Figure 8. Percentage of total biomass by size-class and depth-strata (Upper slope: 400-800 m; Middle slope: 800-1400; Lower slope: 1400-2000) of deep sea fish assemblages in the western Mediterranean (BB: Balearic sub-basin off north-eastern Iberian Peninsula; AB: Algerian sub-basin southern Balearic Islands) and North Atlantic (RT: Rockall Trough; PS: Porcupine Seabight).

3.2. Crustaceans

Decapod crustaceans are one of the dominant megafaunal groups in the Mediterranean deep-sea communities (Abelló et al., 1988; Cartes and Sardà, 1992; Sardà et al., 1994), with increasing relative abundance with depth compared to fishes (Company et al., 2004; Guijarro et al., 2011). The oligotrophic nature of the Mediterranean waters has been presented as one of the main environmental factors contributing to the high abundance of decapod crustaceans in comparison with other oceans, in which other megafaunal invertebrates, chiefly echinoderms, predominate (Tyler and Zibrowius, 1992; Sardà et al., 1994). Despite this, decapod abundance, biomass and diversity decrease with depth in the deep-sea Mediterranean and its biomass is higher in the western basin than in the central and eastern basins (Company et al., 2004).

Although nearly the same species of deep-sea decapod crustaceans are found all across the Mediterranean (Figure 9), the species relative abundance and biomass by area and depth interval is not homogeneous. For instance, *Aristeus antennatus* and *Aristaeomorpha foliacea* are found along the Mediterranean, but its importance differs among areas: the former shows higher abundances in the western and central basins, where it is an extensively exploited species, while the latter species can be only found in fewer areas (Mytilineou et al., 2006). Other species showing geographical differences in abundance are *Nematocarcinus exilis* and *Acanthephyra eximia*, being the former more abundant in the eastern and the later in the western Mediterranean (Company et al., 2004). Similarly, *Parapenaeus longirostris* is more abundant in the eastern and central basin than in the western basin (Abelló et al., 2002). These differences may be related with a general decrease in productivity from west to east and with depth, although fishing activity has also been considered as a factor that could affect individual characteristics as mean size, maximum size and population size structure (Mytilineou et al., 2001).

Different depth ranges for species have been attributed to different levels of primary production observed along the west to east axis in the Mediterranean (Maynou and Cartes, 2000).

The highest number of decapod species has been identified in the western Mediterranean, during bottom trawl surveys carried out between 200 and 800 m depth in the Spanish coasts, both in the mainland (74 species from 245 samples, Abelló et al., 2002) and in the Balearic Islands (75 species from 153 samples, Guijarro et al., 2011).

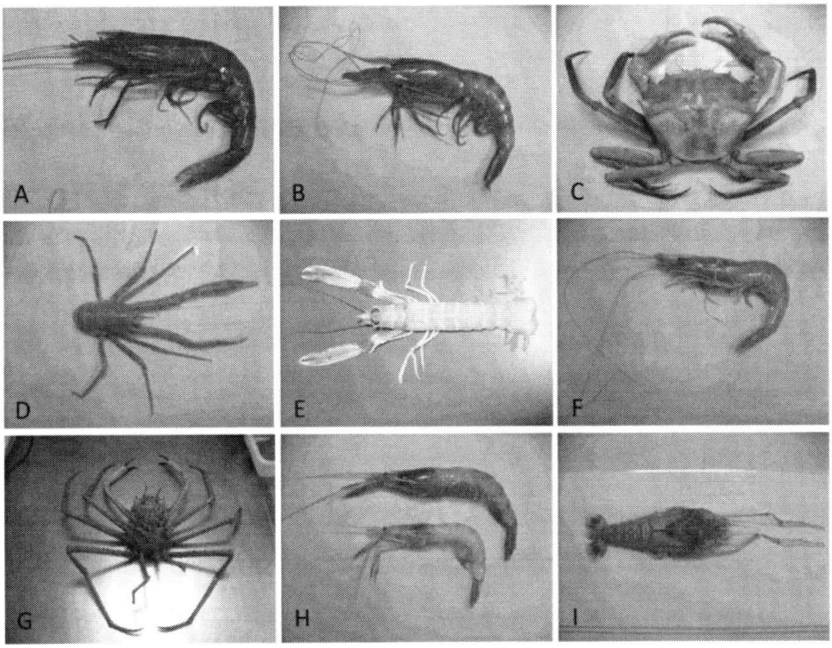

Figure 9. Pictures of some deep-sea crustacean species from the Mediterranean Sea: *Aristaeomorpha foliacea* (A); *Aristeus antennatus* (B); *Geryon longipes* (C); *Munida rugosa* (D); *Nephrops norvegicus* (E); *Parapenaeus longirostris* (F); *Paromola cuvieri* (G); *Pasiphaea multidentata* (H); and *Polycheles typhlops* (I).

Other studies carried out in the western basin between 275 and 640 m depth, reported 51 and 43 species in the Catalan coast (Abelló et al., 1988; Maynou et al., 1996) and 35 species in the Alboran Sea (16 samples; Abad et al., 2007). Fifteen species were reported in 25 hauls carried out both during day and night between 160 and 600 m depth in a submarine canyon in Sardinia (central Mediterranean; Sabatini et al., 2007) and 43 species were reported in the western coast of central Italy between 200 and 700 m depth (Fanelli et al., 2007). Studies carried out in deeper waters showed lower diversities: 23 species identified in 32 hauls between 793 and 1598 m depth in the south-eastern Sardinian waters (Follesa et al., 2009), 39 species in the Eastern Ionian Sea, between 300 and 1200 m depth (Politou et al., 2005) and 32 species identified between 600 and 4000 m depth in the Balearic and Ionian Seas (Company et al., 2004; Tecchio et al., 2011). Table 3 summarises the list of deep-sea crustaceans species caught in the Mediterranean.

Decapod crustacean populations show different bathymetric trends, with an increasing or decreasing trend in mean size with depth depending on the species. Thus, juveniles of *A. antennatus* (Sardà et al., 2004), *Geryon longipes* (Company 1995; Koukouras et al., 2000) and *Polycheles typhlops* (Abelló and Cartes, 1992; Company, 1995) are more abundant in their corresponding deepest distribution ranges, but for other species such as *A. foliacea* (Politou et al., 2004; Kapiris and Kavvadas, 2009), *Parapenaeus longirostris* (Guijarro et al., 2009) and several species of the genus *Plesionika* (Puig et al., 2001; Maiorano et al., 2002; Politou et al., 2000; Guijarro et al., 2012), juveniles concentrate in their shallowest distribution ranges. Sex-ratio of deep-sea decapods is generally significantly biased in favour to one sex, probably as a gender-specific survivorship, although it has also been related to trophic and reproductive behaviour (Company and Sardà, 1997).

Thus, *A. antennatus* is dominated by females in the shallower waters, attaining 50% at certain depths and even showing higher values in the deepest distribution and showing seasonal variations (Sardà et al., 2003; Guijarro et al., 2008).

Table 3. Epibenthic crustaceans collected in the deep waters of the Mediterranean

Species		
Acanthephyra eximia	*Ligur ensiferus*	*Philocheras echinulatus*
Acanthephyra pelagica	*Liocarcinus depurator*	*Plesionika acanthonotus*
Aegaeon lacazei	*Lophogaster typicus*	*Plesionika antigai*
Alpheus glaber	*Macropipus tuberculatus*	*Plesionika edwardsii*
Alpheus platydactylus	*Macropodia longipes*	*Plesionika gigliolii*
Anapagurus laevis	*Macropodia longirostris*	*Plesionika heterocarpus*
Aristeomorpha foliacea	*Maja squinado*	*Plesionika martia*
Aristeus antennatus	*Medorippe lanata*	*Plesionika narval*
Bathynectes maravigna	*Meganyctiphanes norvegica*	*Polybius henslowi*
Calappa granulata	*Monodaeus couchii*	*Polycheles sculptus*
Calocarides coronatus	*Munida intermedia*	*Polycheles typhlops*
Calocaris macandreae	*Munida iris*	*Pontocaris lacazei*
Chaceon mediterraneus	*Munida rutllanti*	*Pontophilus norvegicus*
Chlorotocus crassicornis	*Munida tenuimana*	*Pontophilus spinosus*
Cymonomus granulatus	*Natatolana borealis*	*Processa canaliculata*
Dardanus arrosor	*Nematocarcinus exilis*	*Processa edulis*

Species		
Dorhynchus thomsoni	*Nephrops norvegicus*	*Processa nouveli*
Ebalia deshayesi	*Pagurus alatus*	*Richardina fredericii*
Ebalia nux	*Pagurus excavatus*	*Rissoides pallidus*
Ergasticus clouei	*Pagurus prideaux*	*Rochinia carpenteri*
Euchirograpsus liguricus	*Palinurus mauritanicus*	*Scalpellum scalpellum*
Funchalia woodwardi	*Pandalina profunda*	*Scyllarus arctus*
Gennadas elegans	*Parapenaeus longirostris*	*Sergestes arachnipodus*
Geryon longipes	*Paromola cuvieri*	*Sergestes arcticus*
Goneplax rhomboides	*Parthenope macrochelos*	*Sergia robusta*
Homola barbata	*Pasiphaea multidentata*	*Solenocera membranacea*
Hymenopenaeus debilis	*Pasiphaea sivado*	*Stereomastis sculpta*
Inachus communissimus	*Penaeopsis serrata*	*Squilla mantis*
Inachus dorsettensis	*Periclimenes granulatus*	*Xantho pilipes*

From Abelló et al. (1988; 2002), Company et al. (2004), Abad et al. (2007), Fanelli et al. (2007), Follesa et al. (2009) and Tecchio et al. (2011).

Even for the species of the same genus, such as *Plesionika*, there is not a clear trend, as different authors have detected different trends in the distribution of sexes by depth (e.g. Company and Sardà, 1997; Carbonell and Abelló, 1998; García-Rodríguez et al., 2000; Colloca, 2002; Maiorano et al., 2002; Chilari et al., 2005; Vafidis et al., 2008; Guijarro et al., 2012).

Most species of pandalid shrimps in shallow waters are protandrous hermaphrodites, changing from male to females as they grow. However no examples have been found of deep-sea pandalid shrimps being fully dioecious, although the reason for this difference between deep-sea and shallow species is not known (Young, 2003). Reproductive periodicity of deep-sea crustaceans seems to follow the initial hypothesis of a seasonal breeding pattern (Wolf, 1962) followed by a fluctuating pattern of recruitment. Coastal and shelf dwelling species present highly variable reproductive traits in terms of seasonality, with many species reproducing throughout the year and many other being distinctly seasonal (Giangrande et al., 1994).

The same variability of reproductive period duration has been described for deep-sea species dwelling below 1000 m depth (Tyler et al. 1994). Several deep-water crustacean species showed a marked seasonality in reproductive periods, like *A. antennatus* (Carbonell et al., 1999; Sardà et al., 2003; Guijarro et al., 2008), *Pasiphaea multidentata* (Ramírez-Llodrà et al., 2007) or *P. longirostris* (Guijarro et al., 2009), while others do not show any seasonal trend, such as the case of *Polycheles typhlops* (Follesa et al., 2007) or *Nephrops norvegicus* (Rotllant et al., 2005).

Increasing seasonality in reproductive periods by depth have been found for several species of Pandalidae: from the shallowest species *P. heterocarpus* with ovigerous females present throughout the year, to the deepest species, *Plesionika acanthonotus* with ovigerous females present only in late spring and early summer months (Company and Sardà, 1997).

Furthermore, the analysis of 19 decapod species in the northwestern Mediterranean showed the reproductive activity centred on different seasons, depending on the genus, but also an increasing seasonality of the reproductive periods with increasing species depth distribution (Company et al., 2003). These authors proposed a theoretical relationship between the duration of the reproductive periods and depth of distribution (Figure 10). This progressive decrease in the duration of the reproductive periods has been related to a decrease of matter input, preventing females from allocating energy to their reproductive processes throughout the year.

Deep-water crustaceans have an important commercial interest in the Mediterranean, with high economic values. Figure 11 shows the relative importance of each species caught by the bottom trawl fleet of Mallorca (Balearic Islands), both in terms of biomass and economic value. The red shrimp *Aristeus antennatus* (Fig. 12) is one of the most valuable deep water resources in the western Mediterranean (Carbonell et al., 1999; Sardà et al., 2003), being the target species of a well-developed bottom-trawl fishery between 500 and 800 m depth. It is also exploited in the central Mediterranean, together with *Aristaeomorpha foliacea* (Sardà et al., 2004). Both species are practically unexploited in the eastern basin (Papaconstantinou & Kapiris, 2001) and a recent bio-economic simulation analysis showed than the fishing effort targeting this species could be increased by 50-100%, increasing the fleet's profitability without jeopardizing the sustainability of the fishery (Guillen et al., 2012). Other commercially important species is the Norway lobster *Nephrops norvegicus*, a target species in fisheries operating around 400 m along the Mediterranean (Sardà, 1998). *Parapenaeus longirostris* is especially important in the central Mediterranean, although it can also be considered as a by-catch species in the western (Sbrana et al., 2006).

Other species less important in terms of economic value and biomass for the bottom trawl fishery are pandalid species of the genus *Plesionika*, which are regularly marketed mixed under a common name (Vafidis et al., 2005). A trap fishery targeting pandalid shrimps has also been developed in the

Mediterranean (Thessalou-Legaki, 1989; García-Rodríguez et al., 2000). Other crustacean species caught as by-catch are *Pasiphaea* spp., *Geryon longipes*, *Macropipus tuberculatus* and *Paromola cuvieri* (Guijarro and Massutí, 2006).

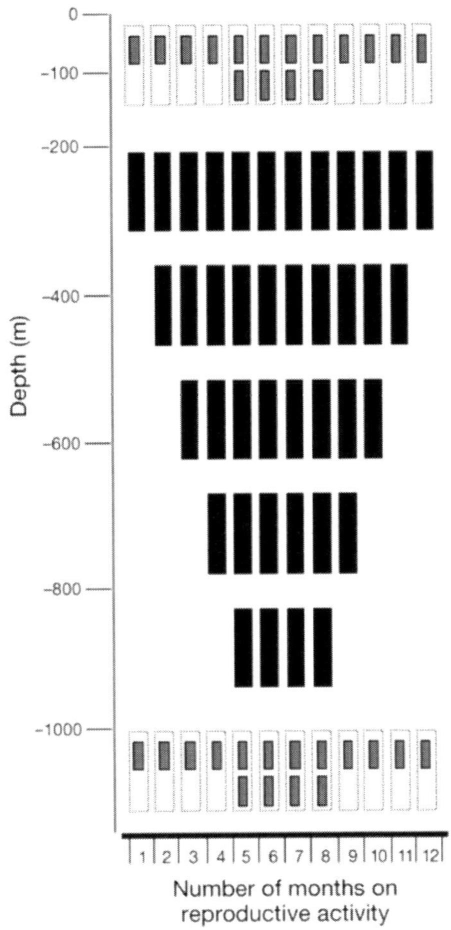

From Company et al. (2003).

Black bars: hypothesised number of months in which species may undertake their reproductive activity supported by the data presented in this study; grey bars: hypothesised co-occurrence of seasonal and continuous reproductive species at shelf and lower-slope depths.

Figure 10. Scheme of the relationship between duration of reproductive periods and depth of distribution of visual-feeder megafauna dwelling across the continental margins of the NW Mediterranean Sea.

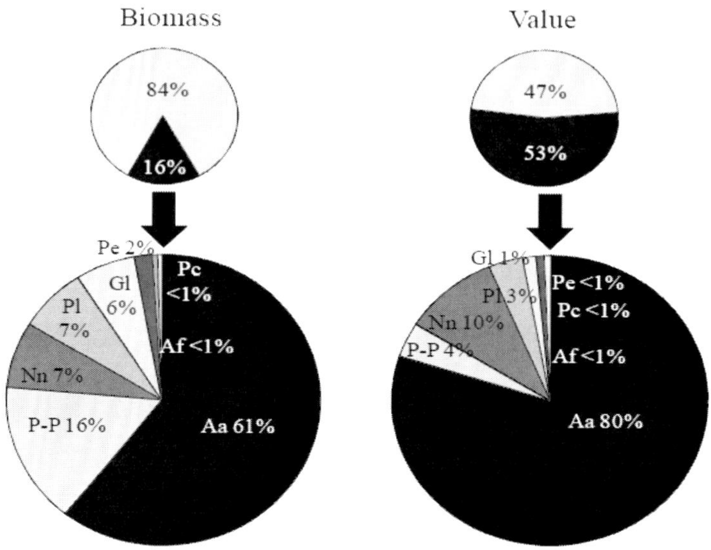

Mean values for 2000-2009. Aa: *Aristeus antennatus*; P-P: Pandalidae and Pasiphaeidae; Pl: *Parapenaeus longirostris*; Gl: *Geryon longipes*; Nn: *Nephrops norvegicus*; Pe: *Plesionika edwardsii*; Pc: *Paromola cuvieri*; Af: *Aristaeomorpha foliacea*. From Guijarro (2012).

Figure 11. Percentage of deep water crustaceans in relation to total landings caught by the bottom trawl fleet of Mallorca (Balearic Islands) (up) by biomass and values and percentage by species (down).

Figure 12. The red shrimp *Aristeus antennatus* is the most important commercial resource for the bottom trawl fishery developed at deep-sea waters in the western Mediterranean. The fishery exploits middle slope grounds down to approximately 800 m depth and takes a large variety of fishes, crustaceans and cephalopods inhabiting the deep Mediterranean.

3.3. Cephalopods

There are estimated to be over 1000 species of cephalopods living in the world today (Norman, 2003).

Although deep-sea cephalopods account for about 85% of all current genera, they are relatively unknown compared to their shelf-living relatives (Clarke, 1996). As occurs in all other taxonomical groups, most information concerning the Mediterranean cephalopod species and assemblages refer to water depths down to about 800 m, that is, the bathymetric range exploited by the bottom trawl fishery.

Deep-sea cephalopods have a series of morphological and histological adaptations for living in such deep, cold, twilight waters. Morphological characteristics of deep-sea octopods were reviewed by Voss (1988), who considered the following main adaptations: 1) loss of the ink sac, in contrast with shallow-water octopuses, this organ is small, degenerate, or absent in deep-sea species because in the absence of light an ink jet does not work as a defense mechanism; 2) reduction of the crop, radula and posterior salivary glands, which is probably related to the change to small, soft-bodied prey; 3) reduction of the gills, in response to decreased metabolic activity in the cold deep-sea waters; 4) large eggs, which are presumably associated with crawl-away young with no, or a very short, planktonic stage, a factor that limits their geographical distribution; 5) subgelatinous tissues, with increasing depth, the mantles are increasingly modified because of the decrease in musculature and the appearance of a gelatinous material that is probably a mean of reducing metabolic requirements.

All these morphological changes, among others such as the presence of few but large spermatophores, the enlarged ligula from the hectocotylus arm or the reduction of the sucker rows or sucker sizes, are evolutionary adaptations to living in deep-water habitats, since it is thought that deep-sea octopuses are derived from their shallow-water relatives (Voss, 1967).

Although very little is known about the physiological adaptations of cephalopods to the deep sea (Seibel et al., 1997), there is a clear trend of decreasing metabolic rates with increasing depth in oceanic species which is believed to result from decreased locomotory capacity for visually cued predator/prey interactions in the low ambient light levels of the deep sea (Seibel et al., 1997; Seibel and Childress, 2000).

However, this do not apply to benthic octopods which have greater opportunities for crypsis and refuge, and consequently have similar metabolic rates regardless of habitat depth (Seibel and Childress, 2000). Cephalopods belonging to the Order Octopoda, which include the Cirrata and Incirrata forms, are abundant in demersal deep-sea communities, whereas the incirrates (Collins and Villanueva, 2006) prevail on the upper slope and cirrates on the lower slope and abyssal plains (Allcock et al., 2003).

The bulk of the Mediterranean teuthofauna comes from the Atlantic Ocean (Bello, 2003). Up to date, a total of 66 cephalopod species has been reported (Bello, 2008) but only 53 of them are represented by well established populations (Bello, 2003). According to Bello (2003), seven cephalopods are endemic and quasi-endemic Mediterranean species, which account for 10.4% of its teuthofauna. Four Lessepsian species have been reported during the last decades: *Octopus aegina* (Salman et al., 1999), *Sepioteuthis lessoniana* (Salman and Katagan, 2002), *Sepia pharaonis* (Mienis, 2003) and *Tremoctopus gracilis* (Orsi-Relini et al., 2004).

The number of deep-sea species (Figure 13), that is, those living below the edge of the continental shelf or deeper than approximately 200 m depth, is rather low compared to those of fishes and crustaceans.

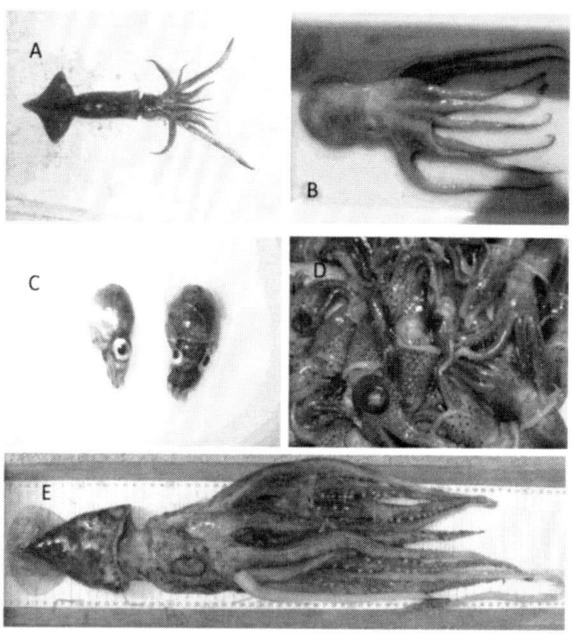

Figure 13. Pictures of some deep-sea cephalopod species from the Mediterranean Sea: the ommastrephid squid *Todarodes sagittatus* (A); the small bathyal octopus *Bathypolypus sponsalis* (B); the sepiolid *Heteroteuthis dispar* (C); and the squids *Histioteuthis reversa* (D) and *H. bonnellii* (E).

Table 4. Frequency of appearance (*f*, number of hauls in which the species was caught in relation to the number of hauls taken at each depth stratum), average catch per hour (g/h) and percentage of biomass of cephalopod species taken by commercial bottom trawlers between 200 and 800 m depth in the western Mediterranean

		200–600	m		600–800	m
Species	*f*	*g/h*	*%*	*f*	*g/h*	*%*
Octopus vulgaris	11.8	117.71	4.77	–	–	–
Sepia elegans	17.6	9.68	0.29	–	–	–
Alloteuthis media	23.5	15.61	0.59	–	–	–
Sepia orbignyana	29.4	11.55	0.43	–	–	–
Loligo forbesi	52.9	231.36	8.51	–	–	–
Eledone cirrhosa	82.4	424.01	18.71	8.7	12.80	3.35
llex. coindetii	41.2	26.96	1.33	–	–	–
Sepietta oweniana	47.1	8.29	0.43	–	–	–
Rossia macrosoma	17.6	7.40	0.34	–	–	–
Scaeurgus unicirrhus	35.3	61.23	2.37	4.3	5.60	0.67
Pteroctopus tetracirrhus	52.9	54.81	3.43	8.7	6.78	1.81
Todadrodes sagittatus	88.2	924.63	41.45	91.3	258.94	59.33
Todaropsis eblanae	11.8	6.06	0.32	–	–	–
Octopus salutii	52.9	199.64	8.30	4.3	2.03	0.28
Neorossia caroli	47.1	33.31	1.43	17.4	2.09	0.57
Bathypolypus sponsalis	52.9	90.38	4.65	69.6	28.55	6.35
Histioteuthis bonnellii	29.4	43.06	2.06	39.1	52.99	12.13
Histioteuthis reversa	29.4	4.67	0.36	78.3	70.37	13.34
Ancistroteuthis lichtensteinii	5.9	8.33	0.21	34.8	2.46	0.62
Abralia veranyi	5.9	0.17	0.01	21.7	1.84	0.40
Onychoteuthis banksii	–	–	–	21.7	1.64	0.32
Brachioteuthis riisei	–	–	–	13.0	1.12	0.14
Chiroteuthis veranyi	–	–	–	13.0	2.02	0.27
Chtenopteryx sicula	–	–	–	13.0	1.30	0.22
Heteroteuthis dispar	–	–	–	13.0	0.25	0.06
Abraliopsis morisii	–	–	–	4.3	0.02	0.01
Opisthoteuthis calypso	–	–	–	4.3	0.59	0.14

The species are arrangend in bathymetric order. From Quetglas et al. (2000).

Sampling carried out between 200 and 900 m off the Balearic Islands using both bottom trawl and mid-water trawl gears yielded a total of 26 cephalopod species (Quetglas et al., 2013a). This number is in agreement with the 27 cephalopods taken between 200 and 800 m depth during a sampling period of 14 months (N=79 hauls) in the same study area (Table 4) using exclusively bottom trawl gears (Quetglas et al., 2000).

Despite the depth ranges are not fully comparable, both estimates are markedly higher than the 15 species caught at 450-2200 m depth during different samplings (N=111 hauls) conducted in the adjacent Catalan Sea between 1985 and 2010 (Fanelli et al., 2009). Other studies carried out in the Catalan Sea reported 9 species between 200-550 m (Sánchez et al., 1998) and 13 species (N=105 hauls) between 200 and 832 m (Sánchez, 1986). These results indicate that cephalopod diversity is higher in waters around the Balearic Archipelago than in the adjacent Catalan Sea, which could be due to differences in biogeographic factors or to the lower fishing intensity in the archipelago than along the peninsular coast (Quetglas et al., 2012).

In the eastern Mediterranean basin, Lefkaditou et al. (2003) collected 26 cephalopods sampling from 300 to 1200 m depth in the Ionian Sea, whereas Salman et al. (1997) took 20 species at depths between 200 and 500 m in the Aegean Sea. The scarce information referring to depths greater than 800 m throughout the Mediterranean indicates that the abundance of cephalopods reduces considerably (Lefkaditou et al., 2003). Indeed, Villanueva (1992) only found 10 species sampling at depths between 1000 and 2000 m in the western basin.

Little is known on the bathymetric patterns of cephalopod diversity and all currently available information only refers to the western basin. Analyzing cephalopod assemblages taken by bottom trawlers from 50 to 800 m depth, Quetglas et al. (2000) and González and Sánchez (2002) found a common trend of higher species diversity on the shelf-break and upper slope than on the coastal shelf and middle slope. However, Fanelli et al. (2012) reported an increase of diversity with depth in samplings carried out from 450 to 2200 m depth; these authors also observed seasonal changes of diversity, with the highest values in spring and summer and the lowest in winter. Sampling along the water column considering different layers (surface, deep scattering layer, benthic boundary layer and bottom), Quetglas et al. (2013b) found an increasing trend of diversity with depth.

There are up to nine incirrate benthic octopods in the Mediterranean Sea, of which only three, *Octopus salutii* (Quetglas et al., 2005), *Pteroctopus tetracirrhus* (Quetglas et al., 2009) and *Bathypolypus sponsalis* (Quetglas et al., 2001; Cuccu et al., 2011), can be considered deep-sea forms based on their bathymetric distribution and their mean depth of occurrence, which falls well below 200 m depth (González and Sánchez, 2002). Of the three deep-sea forms, *B. sponsalis* and *P. tetracirrhus* are two small-sized species that are externally very similar and that share many of the traits typical of the deep-water inhabitants cited above.

Both species have flaccid, gelatinous bodies, relatively large eyes, relatively small suckers, a reduced ink sac (*P. tetracirrhus*) or no ink sac (*B. sponsalis*), a moderately (*P. tetracirrhus*) or highly (*B. sponsalis*) enlarged ligula, and females that lay a small number of eggs that are of medium (*P. tetracirrhus*) or large (*B. sponsalis*) size in comparison with coastal octopuses (Mangold-Wirz, 1973). From the differences between these two species, it is clear that *B. sponsalis* has characteristics that are more adapted to life in the deep sea. This can be explained by the fact that although *B. sponsalis* has been caught at depths as deep as 1835 m (Villanueva, 1992), the deepest limit of *P. terracirrhus* seems to be at approximately 750 m depth (González and Sánchez, 2002). In fact, *B. sponsalis* was the only incirrate octopod captured in samples taken between 1000 and 2000 m depth in the western Mediterranean basin (Villanueva, 1992).

However, *O. salutii* has morphological characteristics of both the deep-sea and coastal forms, since it has a muscular but more flaccid mantle than its shallow-water relatives, well-developed suckers but a very large ligula, and the females lay a moderate number (2000 to 4000 according to Mangold-Wirz et al., 1976) of medium-sized eggs (e.g. larger than *Octopus vulgaris* but smaller than *P. tetracirrhus;* Mangold-Wirz, 1973).

Considering deep-sea squids, the most frequently caught species in the Mediterranean include the ommastrephid *Todarodes sagittatus* (Quetglas et al., 1998, 1999) and the two representatives of the family Histioteuthidae (*Histioteuthis reversa* and *H. bonnellii*) (Bello, 2000; Quetglas et al., 2010). These three species are voracious predators of fishes and crustaceans undertaking diel vertical movements and thus playing an important role in the slope trophic webs from the Mediterranean.

All other deep-sea cephalopod species are caught in very low numbers and indeed the knowledge on its biology and ecology is rather limited and fragmentary. A recent work reviewed the information of some of these species in the Mediterranean (Quetglas et al., 2013a), including one deep-sea cirrate octopus (*Opisthoteuthis calypso*), the sepiolid *Stoloteuthis leucoptera* and seven teuthoid species inhabiting preferentially mesopelagic and bathypelagic waters (*Abraliopsis morisii, Ancistrocheirus lesueurii, Brachioteuthis riisei, Chiroteuthis veranyi, Chtenopteryx sicula, Onychoteuthis banksii* and *Taonius pavo*). With the only exception of *T. sagittatus* (Quetglas et al., 1998), the rest of deep-sea cephalopods do not have commercial interest in the Mediterranean owing to different factors such as low abundance, small size or gelatinous flesh (Sartor et al., 1998).

3.4. Cold Water Corals

The Mediterranean Sea, well known, sailed, fished, trawled and exploited since centuries, has been a complete unknown since the last 70 years in its deep grounds, where hidden Cold-water corals (CWC) living communities inhabit areas located more than 700 m deep in some locations (Freiwald et al., 2009). The first scientific studies published on CWC fauna of the Mediterranean continental shelves were carried out by Pérès and Piccard (1964) who documented the communities of white corals from Mediterranean circalittoral zones. Almost a decade later Desbruyères (1972-73) published its benthic bionomie of the Spanish Catalan coast and in the same years Reyss (1972-73) published his work on the Lacaze-Duthiers and Cap de Creus canyons, which are two of the submarine canyons conforming a canyon system in the Gulf of Lions. The most comprehensive taxonomical work on Mediterranean scleractinians, which include also deep sea species, is the study by Zibrowius (1980), but there are also several published works dealing with the taxonomy of other groups (as gorgonians: Rossi 1959, Carpine and Grasshoff 1975, Weinberg 1976; and black corals: Lapian et al., 2007, Bo 2008). However, a comprehensive taxonomical guide for Mediterranean CWCs has not been compiled yet. A publication from Aguilar (2007) summarizes general information on Mediterranean corals, from shallow to deep areas. From the seventeen's to nowadays important advances in the knowledge on the Mediterranean CWC fauna have been achieved, especially in the last 20 years. Here we try to make a brief overview through some of the aspects of the biology and ecology of the Mediterranean CWCs that we currently know on these for many years' undiscovered communities.

The topographical characteristics of the locations where CWC have been described up to date in Mediterranean waters differ from the sites in the North Atlantic where the best studied CWC occur. Whereas in the Atlantic CWCs develop in very different topographies and substrates (from flat areas to mounds including canyons, and from sandy bottoms to rocky walls), Mediterranean CWCs seem to settle preferentially in steeply inclined walls and gently sloping shelves. This has been documented by Freiwald et al. (2009) for the Central Mediterranean, and seems to agree with most of the records of CWCs in the Alboran Sea (e.g. Templado et al., 1986; Lo Iacono et al., 2008), and was also observed in some of the submarine canyons of the Gulf of Lions, notably Cap de Creus (Orejas et al. 2009; Lo Iacono et al., 2012) and Lacaze-Duthiers (Gori et al., 2013). In Freiwald et al. (2009), several areas of the continental shelf in the Central Mediterranean were documented: the Sicilian channel, the Adriatic basin and the Apulian CWC reefs off Santa Maria di Leuca, the latter is one of the first described locations harboring well developed CWC communities, dominated by *Madrepora oculata* (e.g. Mastrotaro et al., 2001; Tursi et al., 2004; Vertino et al., 2010).

The depth range of those Mediterranean findings ranges from around 150 meters for the shallowest communities in the shelf of the Cap de Creus canyon, to the documented ca. 700 m in the Adriatic (Freiwald et al., 2009). Occurrence of colonial scleractinian CWC displayed in figure 14, reflects the current knowledge on its distribution for the Mediterranean.

However, it is worthy to mention that probably this image would change completely in the next years, as the increased discovery of new sites within the last 10 years has been notable. Indeed, technological advances in acoustic methods allowed to highly increase our knowledge on Mediterranean CWC habitats, and in the last three years the first habitat maps for Mediterranean CWC communities have been published (Santa Maria di Leuca: Vertino et al., 2010, Savini et al., 2014; and Cap de Creus canyon: Lo Iacono et al., 2012).

The dominant scleractinian colonial species in the currently investigated Mediterranean waters are *Madrepora oculata*, *Lophelia pertusa* and *Dendrophyllia cornigera*, whereas among the solitary species, the cup coral *Desmophyllum dianthus* is the most frequent one (Figure 15).

The dominance of *M. oculata* and *L. pertusa* has been already documented by several authors (e.g. Pérès and Picard 1964; Freiwald et al., 2009; Orejas et al., 2009; Gori et al., 2013).

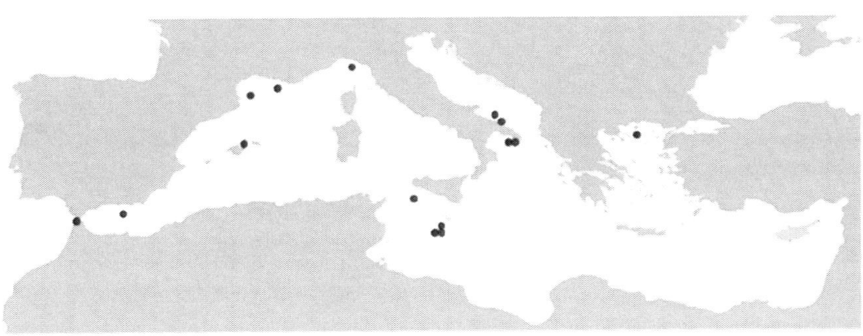

Source: Aguilar (2007), Freiwald et al. (2009) (and references therein), Orejas et al. (2009), Vertino et al. (2010), Gori et al. (2013).

Figure 14. Occurrence of live *Madrepora oculata* and/or *Lophelia pertusa* across the Mediterranean Sea.

Deep sea megabenthic communities dominated by gorgonians seem to be relatively scarce in Mediterranean waters however; the bamboo coral *Isidella elongata* constitutes an example of them (e.g. Pérès, 1967; Bellan-Santini, 1985; Laubier and Ewig, 1993).

Among the sea pens, *Funiculina quadrangularis* is a notable case of a dominant species in soft substrates (Maurin, 1968; Vafidis et al., 1994; Nouar and Maurin, 2001). Very important in several places are the communities dominated by Anthipatarians (black corals) which have been documented in several areas in Mediterranean waters, as for instance in the Menorca Channel and the Central Mediterranean (Bo et al., 2009, 2011, 2012) (Figure 16).

Image A ©A. Gori. Images B, D ©E. Obis. Image C ©A. Olariaga.

Figure 15. Colonial and solitary cold-water coral species most abundant in the Mediterranean. A. *Madrepora oculata*, B. *Lophelia pertusa*, C. *Desmophyllum dianthus*, D. *Dendrophyllia cornigera*.

The presence of CWC communities dominated by scleractinians or gorgonians is also known from the Alborán Sea (Aguilar, 2007; Lo Iacono et al., 2008), and Mediterranean seamounts as Emile Baudot off the Balearic

Islands (Marin et al., 2011) (Figure 17). The fossil record (e.g. Titschack and Freiwald, 2005; Taviani et al., 2005b and references therein) evidences that CWC communities had better times in the past than in current times. Indeed from the study of Zibrowius (1980) it was concluded that *L. pertusa, M. oculata* and *D. cornigera* are in recession in the Mediterranean Sea. One of the most extended hypotheses for the recession of those species, based on the current patched occurrence of CWC in the Mediterranean, is the recolonisation of the Mediterranean shelf by those organisms after the Messinian salinity crises (Taviani, 2002). Not only CWC present evidence for this re-colonization after the Messinian crises, other groups as foraminiferans (Wright, 1978), or bryozoans (Harmelin and d'Hondt, 1992) seems to exhibit the same pattern. The period before, the Mediterranean CWC communities dramatically diminished since the postglacial period (Delibrias and Taviani, 1985). Based on ^{14}C dating, these authors located the major decline of CWC abundance in the Mediterranean Sea from 30 to 15 kyr BP the, whereas other researchers have suggested different dates: McCulloch et al. (2006) dated the demise of CWCs in the Mediterranean to the end of the Younger Dryas (10.5 kyr BP). Following this hypothesis, the expansion of these CWC after the opening of the Gibraltar strait took place from the Pliocene (5 Ma) to Early Pleistocene (1.6 Ma) and from the Western to the Eastern basin (Titschack and Freiwald, 2005). Abundant rests of those past flourishing communities have been found in the fishermen nets across the whole Mediterranean. The possibility to date those dead corals, using new developed geochemical techniques (notably the use of stable isotopes trace and minor elements composition, e.g. Montagna et al., 2005; Eltgroth et al., 2006; Rüggeberg et al., 2008) allow to identify the date of the death of those corals, as well as to use them as climatic archives, contributing to validate the hypothesis. Montagna et al. (2008) presented how trace element concentration combined with U-series and radiocarbon dating of the skeleton of Mediterranean *D. dianthus* and *L. pertusa*, are a valid high-resolution Surface Sea Temperature (SST) archive, as well as seawater chemistry and ocean circulation proxy.

The current dominance of *M. oculata* in the Mediterranean Sea has been reported in some regions, such as the Central Mediterranean (e.g. Taviani et al., 2005a; Freiwald et al., 2009; Vertino et al., 2010), and Western Mediterranean (e.g. Orejas et al., 2009; Pardo et al., 2011; Gori et al. 2013). This seems to be related with its apparent higher tolerance to environmental changes compared to *L. pertusa*, as it has been recorded in a study of the fossil community (Wienberg et al., 2009). Further, Keller and Os'kina (2008) also documented a higher thermal tolerance for *M. oculata*, which is able to survive at temperatures up to 20°C. Common features of the current Mediterranean locations of CWCs are the dominance of hard substrates, high hydrodynamic environments, and temperatures around 12-14°C, reaching a maximum of 15.5°C, as it has been recorded at 300 m water depth for instance in the Cap de Creus Canyon (Palanques et al., 2006).

The relation established by Dullo et al. (2008) between CWC occurrence and hydrography (density envelope sigma-theta (δ_θ)) of the Celtic and Nordic European continental margin would probably not be suitable to explain the occurrence of CWC in the Mediterranean. Hence, the water density at the depths were CWCs live in the Mediterranean is much higher (de Madron et al., 2005).

Traditionally coral reefs, considering the tropical as well as the temperate and cold ones, have been described as communities which have a highly diverse associated fauna. A recent study on some CWC communities of the Northeast Atlantic confirm high α, β and ϒ biodiversity index for those ecosystems (Henry et al., 2010). The biodiversity studies in deep Mediterranean waters, or at least including them, are still scarce (Coll et al., 2010; Danovaro et al., 2010; Bo et al., 2012 and references therein), and especially scarce on CWC habitats (Tursi et al., 2004).

However, recent publications present a high species diversity belonging to different groups, mainly dominated by sponges and corals (e.g. Cerrano et al., 2010; Gili et al., 2011; Bo et al., 2012 and references therein). Up today the work from Pérès and Picard (1964) and by Desbruyères (1972-73), which presents a detailed bionomic map as well as a comprehensive species lists for the Catalan shelf, are the most complete studies published for Mediterranean deep mega benthic communities.

Additionally, other authors offer a view of the diversity of suprabenthic organisms (e.g. Cartes et al., 1998, 2009) which inhabit the same biotopes and are potentially associated to these communities. Several studies as the one from Colloca et al. (2004) or Ordines and Massutí (2009), showed the role of Mediterranean communities dominated by megaepibenthic organisms, enhancing the diversity of demersal species (including fish and invertebrates) as well as nursery. This refuge and nursery function of those 3D habitats is known from other ecosystems dominated by CWCs in other latitudes (e.g. Jonsson et al., 2004; Buhl-Mortensen et al., 2010; Purser et al. 2013 among others). However, up to date no studies have been conducted to explore these aspects in the Mediterranean CWC communities.

Images A-D, ©ICM (CSIC) / IFM-GEOMAR. Images E-H ©Milano Bicocca University (CoNISMa).

Figure 16. Overview of cold-water corals from the Balearic Islands (A-D) and Central Mediterranean (E-H). A. Gorgonian group belonging to the species *Callogorgia verticilata* (240 m depth, Menorca Channel), left of the image and exemplar of *Leiopathes glaberrima* could be observed. B. Isolated colony of the black coral *Antipathes dichotoma* (156 m, Menorca Channel). C. Specimens of the black coral *Leiopathes glaberrima* (240 m depth, Menorca Channenl). D. Gorgonian group of the species *Eunicella* sp (150 m depth, Menorca Channel). E, F, H. Coral Framework habitat in the Central Mediterranean (Vertino et al. 2010) dominated by colonies of *Madrepora oculata*; note the common sponge *Pocillastra compressa* (E) and iso-oriented *Madrepora* colonies (H). G. Boundary between Coral Framework and Hardground Crusts and Boulders habitat (Vertino et al. 2010); note a large colony of *Leiopathes glaberrima* (Santa Maria di Leuca Coral Province, Ionian Sea, E, F: MS04, 630 m; G, H, MS06, ca. 500 m).

Images ©OCEANA.

Figure 17. Cold water coral communities in the Alborán Sea. A. Community dominated by black corals (*Parantipathes larix*) and the sponge *Asconema setubalensis* (375m, Cabliers). B, C. Reefs of *Lophelia pertusa* (320m, 345m, respectively, Cabliers). D. Large specimen of *Lophelia pertusa* (320 m, Cabliers). E. Groups of specimens of the bamboo coral *Isidella elongata* (515 m, Seco de los Olivos). F. large specimen of *Leiopathes glaberrima* surrounded by fishes (230 m, Seco de los Olivos).

As already mentioned, we have now a better knowledge about the characteristics of suitable locations in the Mediterranean for the development of CWCs.

Moreover, thanks to the predictive models and the high quality multibeam bathymetry as well as sidescan sonar profiles, combined with imagery, we currently can predict potential locations for CWC_s elsewhere (e.g. Bryan and Metaxas, 2007; Tittensor et al., 2009; Davies et al., 2011).

However, our knowledge on the distribution patterns of these organisms and the factors which determine them is still in its infancy, and the Mediterranean is not an exception. Freiwald et al. (2009) made a good description of the variability of CWC communities in three different locations of the Central Mediterranean Sea: different communities have been described but also some common characteristics were defined: they are patched and dominated by *M. oculata*.

Up to date, the two Mediterranean CWC living locations which have been investigated more deeply are the Reefs off Santa Maria di Leuca (SML) in the Central Mediterranean (northern Ionian Sea) and the CWCs communities of the Cap de Creus and Lacaze-Duthiers submarine canyons (Gulf of Lions). The works of Tursi et al. (2004), Taviani et al. (2005a), Freiwald et al. (2009) presented, among others, data on biodiversity, as well as a description of the communities dominated by CWCs in the Central Mediterranean. The recent works from Vertino et al. (2010) and Savini et al. (2014) showed the most comprehensive habitat characterization of the SML coral province, displaying the diversity of the area (Figure 16). In all studied locations where CWC appeared, the authors found a clear dominance of *M. oculata*, which seems (considering the current knowledge) to be the most abundant deep reef-forming species in the Mediterranean, compared to the dominance of *L. pertusa* in Atlantic waters. The CWC communities inhabiting the scarped walls of the Gulf of Lions canyons are also dominated by *M. oculata,* as revealed the first habitat mapping for the CWC communities in the Cap de Creus canyon (Lo Iacono et al., 2012). Two recent studies deal with the CWC communities of Cap de Creus and Lacaze-Duthiers canyons (Orejas et al., 2009; Gori et al., 2013, respectively).

These works presented the patched distribution of the corals, mostly located in vertical walls, overhangs and rocky outcrops and displaying maximal density values for *M. oculata* around 11 colonies$\cdot m^{-2}$ for Cap de Creus and 13 colonies$\cdot m^{-2}$ for Lacaze-Duthiers (Orejas et al., 2009; Madurell et al., 2010). Some surprising differences have been detected between both canyons: although *M. oculata* dominates in both features, *L. pertusa* conforms frame work structures in some locations of the Lacaze-Duthiers canyon head reaching large sizes (Gori et al., 2013). Some images representing the CWC communities in both canyons are displayed in Figure 18.

Until the last decade, approaches to determine growth rates of CWCs have been conducted mainly by paleontologists and climate researchers (see Sherwood and Risk, 2007 for a review). The skeleton made from aragonite of those organisms, could be accurately dated by means of mass spectrometric

$230^{Th}/U$ dating (e.g. Mangini et al., 1998; Cheng et al., 2000; Douville et al., 2010). There are several methods (e.g. lithium/magnesium(Li/Mg) molar ratios, radio carbon measurements($D_{14}C$) and neodymium isotopes ($_{e}Nd$)) which show that CWCs have the potential to record (as an archive) and help to reconstruct episodes of the past, as for instance rapid temperature changes (e.g. Montagna et al., 2008), water mass ventilation (e.g., Adkins et al., 1998; Frank et al., 2005; Sherwood et al., 2008) and water mass provenance (e.g. van de Flierdt et al., 2010; Montero-Serrano et al., 2011; López-Correa et al., 2012).

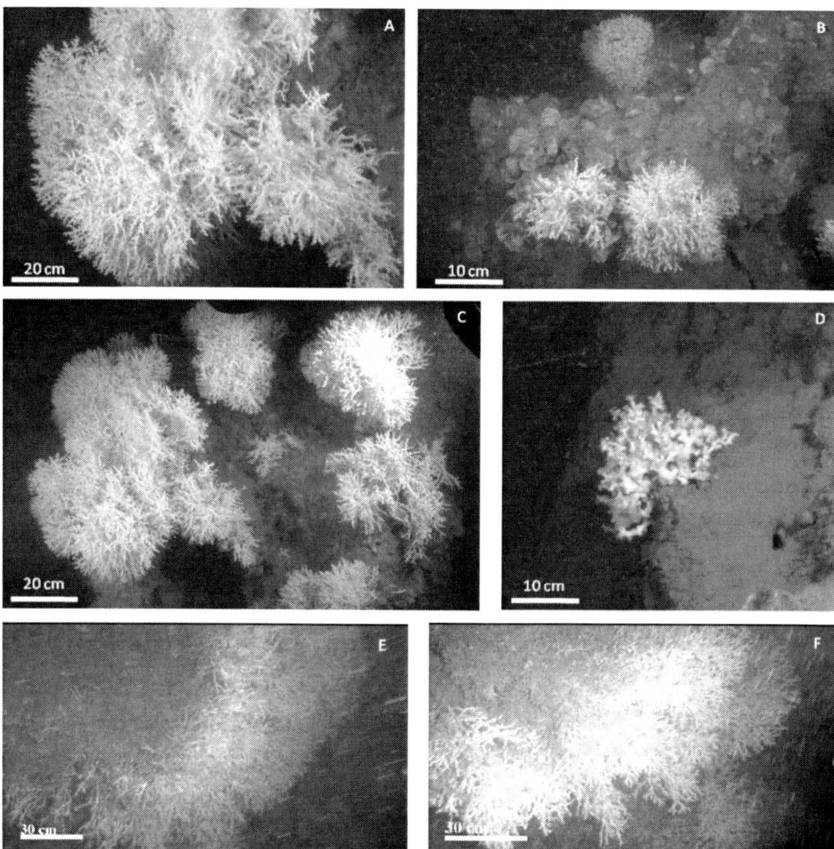

Images A, B, C ©ICM (CSIC) / IFM-GEOMAR. Images D, E ©AAMP.

Figure 18. Overview of Cold-water corals from two canyons of the Gulf of Lions. A, B, C different aspects of the communities dominated by *Madrepora oculata* in the Cap de Creus canyon (200-350 m). D, Specimen of *Lophelia pertusa* in the Cap de Creus canyon (207 m). E,F Coral frame work dominated by *Lophelia pertusa* in the Lacaze Duthier canyon (350 m).

In the last ten years much advances have been done in investigating the growth rates of alive CWCs in different latitudes, notably scleractinians (e.g. Brooke and Young, 2009; Houlbreque et al., 2010; Gass and Roberts, 2011), and bamboo corals (e.g. Andrews et al., 2009; Parrish and Roark, 2009; Thresher, 2009). Several works have also been developed in the Mediterranean, mainly conducting aquaria experiments. These kind of experiments are currently a necessary tool to better understand the functionality of these organisms, as the development of in situ experiments at great depths is complicated, time consuming and very expensive. The good adaptation of many species of CWC (especially scleractinians) to the aquaria conditions, allowed to record photographically the growth of *M. oculata* and *L. pertusa* exemplars (see Figures 1, 2 in Orejas et al., 2008). CWC linear growth rates obtained range from 3-18 mm·y^{-1} for *M. oculata* and 15-17 mm·y^{-1} for *L. pertusa*; these have been the first growth rates obtained for alive CWCs. During the last five years several works dealing with CWC growth measurements applying different methodologies have been published (e.g. Maier et al., 2009, 2011; Orejas et al., 2011a,b). Although the most studied species up to date have been *L. pertusa* and *M. oculata,* some data is available for other species, as *Dendrophyllia cornigera* and *Desmophyllum dianthus* (Orejas et al., 2011a). A comprehensive review for CWC growth rates from elsewhere has been summarized in Roberts et al. (2009). Recently French researchers (Lartaud et al., 2013) conducted the first in situ trials in the Lacaze-Duthiers canyon to obtain growth rates for *L. pertusa* and *M. oculata* and simultaneously accomplished aquaria experiments in order to detect possible differences. Interestingly, their results show the lack of statistical significance between the *in situ* and *in aquaria* obtained growth rates, which support the validity and robustness of aquaria measurements to extrapolate to the natural conditions.

One of the aspects completely unknown for Mediterranean CWCs is their reproduction. The studies conducted up to date dealt with specimens from Atlantic waters, especially *L. pertusa* and *Oculina varicosa* (Brooke and Young, 2005; Waller and Tyler, 2005), and further studies of deep solitary cup corals have also been conducted by Waller (2005) and Waller et al. (2002, 2011). The reproduction of other groups of CWCs, as gorgonians, black corals or hydrocorals, even in other latitudes are almost complete unknown and just a few papers have been published for these groups (Orejas et al. 2002, 2007; Waller, 2005; Waller et al., 2008).

The feeding of four Mediterranean CWCs (*L. pertusa, M. oculata, D. cornigera* and *D. dianthus*) has also been studied in aquaria (Tsounis et al.,

2010). Both Mediterranean and Atlantic studies, have been developed quantifying prey consumption rates (Tsounis et al. 2010; Purser et al., 2010) and using stoichiometric analyses (Kiriakoulakis et al., 2005; Duineveld et al., 2007; Dodds et al., 2009). The prey capture experiments showed distinct rates for the different species and different efficiencies depending on prey size and current speed. For example, *L. pertusa* showed a higher efficiency feeding zooplankton by low current speeds (around 2.5 $cm \cdot s^{-1}$) than by higher rates. In the work from Dodds et al. (2009), using lipid biomarkers, zooplankton was found to be the main contributor to *Lophelia's* diet, even though exhibiting differences between shallower and deeper populations.

A further contribution, which increase the knowledge on CWC's feeding ecology, has been supplied by Carlier et al. (2009) who presented the first results of stable isotopic analyses with corals from the Mediterranean CWC site of Santa Maria di Leuca. Using this method, the authors detected that *M. oculata* and *L. pertusa*, as well as the gorgonian *Paramuricea* cf. *macrospina*, have a diet mainly composed of zooplankton, whereas the diet of the black coral *Leiopathes glaberrima* seems to be less zooplankton dependent. Trying to better understand the role of the diet in the metabolism of those organisms, ecophysiological studies with the cup coral *D. dianthus* brought us a step further showing how the consumption of zooplankton highly influence some vital processes as respiration and calcification (Naumann et al., 2011).

A couple of years ago, it was still thought that the Mediterranean Sea offers an environment to the CWC, where they survive at the limits of their biological capabilities. Hence, it was assumed that the relative high temperatures of the Mediterranean in the areas where these organisms develop (around 12.5ºC) were the biological thermal limits for those organisms. Especially when taking into account those from the North Atlantic, where these organisms develop reefs, since they inhabit areas where temperatures do not reach the 9ºC even in summer.

However, new ecophysiological experiments discovered that *D. cornigera* and *D. dianthus* grow well under elevated temperatures (17ºC), which correspond to sea surface values. Indeed, *Dendrophyllia* showed higher growth rates by 17ºC which is well above (around 5ºC) the average temperatures of the areas where the species normally develops in Mediterranean sites (Naumann et al., 2013).

Currently, a lot of ecophysiological research is being conducted elsewhere and also in the Mediterranean sea, trying to better understand the ecological limits and capabilities of these species and in order to be able to predict the response of those organisms confronted with environmental changes, as global warming or ocean acidification (OA). In this sense, several experiments are now being developed in order to know how CWC could cope with a pH decrease. First results of short term experiments for Mediterranean CWCs have been published by Maier et al. (2009, 2011), showing that pH decrease in a short term scale (days) mostly affect younger compared to older polyps, which implies that skeletal growth of young corallites could be more negatively influenced by OA. Moreover, the experiments conducted by Maier et al. (2011) revealed that due to the pH diminution, the calcification in Mediterranean CWC has already drastically decreased (by 50%).

In the last years, long term experiments analyzing the potential effects of OA in the growth rates of the CWCs have been started in several geographical areas, also in the Mediterranean (Movilla et al., 2014a,b). The results obtained showed species specific effects of pH decrease, as well as different responses of the corals through time, revealing, for instance, a better acclimation of *D. cornigera* to lower pH values, and a decrease in the growth rates of *D. dianthus* under acidified conditions (Movilla et al., 2014b).

4. Human Impacts

Owing to its long history of human occupation, current coastal overcrowding and inherent geographic characteristics (semi-enclosed sea with narrow connections with the open oceans and large coastal area compared to its volume), the Mediterranean is one of the most impacted seas in the world (Bianchi and Morri, 2000; Lejeusne et al., 2010).

Although human impacts are more evident on coastal waters, there is currently mounting information pointing to important effects on deep sea ecosystems. In studies carried out in the Gulf of Lions, for instance, most debris were collected in canyons and bathyal plains, with high amounts occurring deeper than 500 m (Galgani et al., 1996). Vessel generated refuse is a major source of litter into the Mediterranean marine environment (Galil et al., 1995), including plastic (Figure 19), glass bottles, metallic objects or fishing gear remains (Galil et al., 1995; Galgani et al., 1996). Pollutants have also been found in tissues of deep Mediterranean species, such as dioxins in the commercially important red shrimp *Aristeus antennatus* (Rotllant et al., 2006) or organochlorine compounds (PCBs and DDTs) in fishes (Storelli et al., 2007; Storelli and Perrone, 2010). Both the litter and pollutant accumulations on deep sea environments and species highlights the potential threat posed to the biodiversity of such fragile and vulnerable ecosystems (Rotllant et al., 2006).

Other major threats include the fishing exploitation, which progressively moves to deep-sea waters as the shallower resources are being depleted (Danovaro et al., 2010) or, in the case of the Mediterranean, by the high value of deep-sea decapods crustaceans (WWF/UICN, 2004).

Mediterranean deep-sea grounds are almost exclusively exploited by the bottom trawl fishery targeting the decapod crustaceans Norway lobster (*Neprhops norvegicus*) and the aristeid shrimps *Aristeus antennatus* and *Aristaeomorpha foliacea* on the upper and middle slope depths (Moranta et al., 2008). These Mediterranean deep-sea fisheries started during the 1940s/1950s and work on bottoms down to 800 m, rarely reaching the 1000 m depth. Other minor deep-sea fisheries include longliners targeting the European hake *Merluccius merluccius* in the Gulf of Lions and the Ionian Sea or the shark *Hexanchus griseus* in the Aegean Sea (Sardà et al. 2004).

Figure 19. Some examples of the debris collected by the bottom trawl fishery in deep Mediterranean waters, including plastic, glass bottles, metallic objects or fishing gear remains.

There is currently scarce information concerning the effects of trawling on the soft muddy bottoms where this fishery takes place, but the extreme vulnerability and low recovery rates of deep sea beds to physical perturbations suggest that such external disturbances may be very long lasting (Tudela, 2004). Tudela (2004) highlights the injurious effects of otter trawling on the octocorallian *Isidella elongata* facies of the bathyal mud biocenosis. The high amounts of discards from the deep decapod crustacean fishery, which averaged 42% of the total catch in the Balearic Sea (Moranta et al., 2000), gives a comprehensive idea of the trawling impact on the deep-sea megafaunal assemblages. Besides the direct effects on individual species, such high discarded biomass might alter the structure and dynamics of ecosystems through favoring scavengers (Moranta et al., 2000). Currently, there is a ban for trawling at depths greater than 1000 m in the Mediterranean. Although the potential fishing interest of the unexploited grounds below 1000 m depth is very limited, the consequences of fishing on such highly vulnerable habitats could be important not only for the megafaunal communities but also for the harvesting, since reproduction of many commercial species occurs at such great depths (WWF/UICN, 2004).

Contrary to soft muddy bottoms, there is abundant information of human impacts on hard rocky ecosystems such as the cold water coral communities. Besides the obvious effect of bottom trawling, other gears such as the long lines have been also recognized as a danger for these communities (Lumsden et al., 2007), and its impact was observed in Mediterranean areas too (Figure 20). The effect of this anthropogenic impact in the Mediterranean has been recorded and quantified (Freiwald et al. 2009; Orejas et al. 2009).

However, in the last years other direct impacts potentially threat CWC communities: the oil and minerals prospection, so called "drilling" and "mining". The main potential medium term effect from oil prospection is the increase in resuspended sediment due to the drilling activity (Figure 21). CWCs, as filter feeders cannot cope with high sediment load, hence, its filtration mechanisms and structures may result clogged. A high input of sediment could cause the death of the corals. Some experimental work to study the potential effects of this activity in CWC reefs in Norway have already been developed, showing that *L. pertusa* seems to have an efficiently cleaning mechanism (e.g. mucus shedding or eventual cilliary action) for deposited sediment particles smaller than 1mm. Simulating a drilling event, the number of coral fragments with affected tissue and polyp mortality increased, hence the authors concluded that burial of coral by drill cuttings, considering the current threshold level used in environmental risk assessment studies by the offshore industry, may result in damage to *L. pertusa* (Larsson and Purser, 2011). In the Mediterranean Sea, several companies intend currently to start prospection in several locations, for instance in the Gulf of Lions. Up to date no scientific studies have been conducted in the Mediterranean in order to analyze possible effects of the drilling activity on CWC communities.

The increasing knowledge on the occurrence of these communities, simultaneously with advances in the study of its biology and ecology, together with the increasing threats of these ecosystems, urge to take actions in order to apply the gained scientific knowledge in the protection, conservation and proper management of these areas. One of the aspects where more research efforts are needed is the study of the population connectivity (genetic studies), such as the recently published by Dahl et al. (2012) for Swedish waters. This would help to better understand the exchange between populations and apply suitable management measurements.

In the Mediterranean there is currently one CWC protected area: Santa Maria di Leuca (GFCM-FAO 2011) which has been declared as Specially Protected Area of Mediterranean Importance (SPAMI) and has been closed for fishing activities.

Image ©ICM (CSIC) / IFM-GEOMAR.

Figure 20. Colony of *Madrepora oculata* entangled in the remaining of a fishing gear in the Cap de Creus canyon (200 m).

Further, the French Marine Protected Areas Agency has an ongoing project to declare a Marine Natural Park in the Gulf of Lions. Currently, three Mediterranean off shore areas in Spain are studied in frame of the INDEMARES project (LIFE +) (www.indemares.es) in order to be part of the European Natura 2000 network: the Cap de Creus canyon, the Menorca channel and the Seco de los Olivos seamount. All of them harbor CWC communities. The declaration of these areas as MPAs will be a remarkable step enhancing the importance of these areas for the Mediterranean countries.

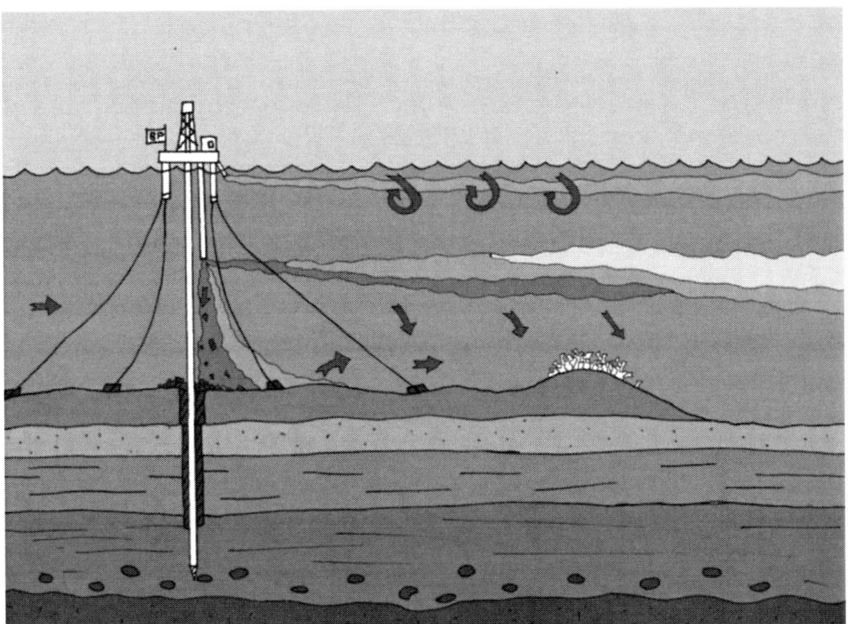

Image ©A. Purser.

Figure 21. Schematic view of a typical drilling operation. The rig is held in place by large anchors (concrete blocks) attached to the rig by cables. Anchor blocks are carefully placed on the seabed to avoid disruption of key communities. The figure shows the release of the drill cuttings (rock chippings) near the surface, at mid water depth, or near the seabed (represented by grey scales). The red arrows indicate how the marine currents can transport the drill cuttings from point of release downstream. The larger cuttings fall to the seabed near the rig, whereas finer drill cuttings can travel considerable distances from the rig, possibly reaching ecosystems such as the coral reef depicted in the figure. The deepest arrows show that after material has settled to the seabed, it can be resuspended during storms or periods of high flow, to possibly be transported to the reef environment at a later date.

Contrary to what could be expected, climate change constitutes a major threat for the biodiversity of Mediterranean deep-sea ecosystems. Given that deep-sea waters are influenced by the sinking of upper layers of increased density (salinity), they will be affected by any major change occurring at the surface. In fact, the first definite sign of temperature changes in deep-sea waters came from the Mediterranean (Lejeusne et al., 2010). Bethoux et al. (1990) reported a general warming trend of 0.12°C analyzing a 30-year time series (1959-1989) of deep-water temperatures from the western Mediterranean, a trend further confirmed by other works (Bethoux et al., 1999; Vargas-Yánez et al., 2008). In addition, major changes in the thermohaline circulation were observed through the 1990s in the eastern Mediterranean, a phenomenon known as the Eastern Mediterranean Transient (EMT). The EMT was characterized by the massive formation of dense, relatively warm water in the Aegean Sea as a result of increased salinity, which eventually caused the uplift of colder, nutrient-rich deep water (Danovaro et al., 2001). This event had even effects on the thermohaline circulation of deep-sea waters from the western basin (Schroeder et al., 2008).

Such climate changes are undoubtedly affecting the marine communities and the life-history of its inhabitants. Although most available information from the Mediterranean comes from shallow waters, deep-sea species could be affected either directly by environmental perturbations of its habitats or indirectly through the benthic-pelagic coupling (Danovaro et al., 2001; Lejeusne et al., 2010). The majority of impacted organisms are sensitive stenothermal species of sponges, cnidarians, bryozoans and mollusks (for a review, see Lejeusne et al., 2010).

The uplift of nutrient-rich deep waters during the EMT increased threefold the primary productivity over the continental shelf and upper slope, causing changes in phytoplankton assemblage composition, species dominance and average phytoplankton cell size (Danovaro et al., 2001). As a consequence of these changes, the biomass of benthic bacteria and meiofauna from the deep Mediterranean waters decreased by as much as 90% and 80%, respectively (Danovaro et al., 1995, 2000, 2001). Using a decadal data set, Danovaro et al. (2004) provided evidence that Mediterranean deep-sea nematode diversity were strongly and rapidly affected by temperature shifts related to the EMT. Once the EMT passed, nematode diversity returned to previous values, providing compelling evidence of the tight connection between environmental changes and deep-sea biodiversity (Danovaro et al., 2004).

The current scarcity of information at worldwide level about the impacts of climate change on deep-sea species and ecosystems is probably related to the difficulties in obtaining data time series at such great depths. In this sense, the results obtained in the Mediterranean Sea during the last decade are thus especially relevant and reinforce the idea of using it as a miniature ocean model to forecast the response of the global oceans to different disturbances. Unexpectedly, such results revealed that deep-sea ecosystems are effectively sensitive and display relatively rapid responses to climate variations occurring hundreds or thousands meters above the sea bottom (Danovaro et al., 2001).

Acknowledgments

Thanks to the European Projects HERMES (Goce-CT-2005-511234-I) and HERMIONE (Grant Agreement Number 226354), the EU Life + project INDEMARES, and the Spanish project DEEP CORAL (CTM2005-07756-C02-02/MAR). We are grateful to the UTM (CSIC) and the crew of the RV "Garcia del Cid" and "Minibex" (COMEX S.A.). We thank the JAGO-team (IFM-GEOMAR, Kiel, Germany) for Cap de Creus canyon and Menorca Channel images. Lacaze-Duthiers canyon images were obtained during the MEDSEACAN cruise, supported by Marine Protected Areas Agency (AAMP, France). We are also grateful to OCEANA for underwater images from Alborán Sea as well as A. Vertino, C. Corselli and G. Gay for the images from the Central Mediterranean CWC, these have been obtained in the frame of the APLABES Programme "CoNISMa APLABES (Apulian PLAteau Bank Ecosystem Study), and a FIRB grant RBAU 01RKC7_004. Thanks to A. Purser (Jacobs University, Bremen, Germany) for providing the illustration of the drilling operation.

References

Abad, E., Preciado, I., Serrano, A., Baro, J., 2007. Demersal and epibenthic assemblages of trawlable grounds in the northern Alboran Sea (western Mediterranean). *Sci. Mar.* 71, 513-524.

Abelló, P., Carbonell, A., Torres, P., 2002. Biogeography of epibenthic crustaceans on the shelf and upper slope off the Iberian Peninsula Mediterranean coasts: implications for the establishment of natural management areas. *Sci. Mar.* 66, 183-198.

Abelló, P., Cartes, J. E., 1992. Population characteristics of the deep-sea lobsters *Polycheles typhlops* and *Stereomastis sculpta* (Decapoda: Polychelidae) in a bathyal mud community of the Mediterranean Sea. *Mar. Biol.* 114, 109-117.

Abelló, P., Valladares, F. J., Castellon, A., 1988. Analysis of the structure of decapod crustacean assemblages off the Catalan Coast (Northwest Mediterranean). *Mar. Biol.* 98(1), 39-49.

Adkins, J. F., Cheng, H., Boyle, E. A., Druffel, E. R. M., Edwards, R. L., 1998. Deep-sea coral evidence for rapid change in ventilation of the deep North Atlantic 15,400 years ago. *Science* 280, 725-728.

Aguilar, R., 2007. *The corals of the Mediterranean*. Oceana, Fundación Ermenegildo Zegna. 94 pp.

Allcock, A. L., Collins, M. A., Vecchione, M., 2003. A redescription of *Graneledone verrucosa*. *J. Moll. Stud.* 69, 135-143.

Andrews, A. H., Stone, R. P., Lundstrom, C. C., DeVogelaere, A. P., 2009. Growth rate and age determination of bamboo corals from the northeastern Pacific Ocean using refined 210Pb dating. *Mar. Ecol. Prog. Ser.* 397, 173–185.

Bellan-Santini, D., 1985. The Mediterranean benthos: reflections and problems raised by a classification of the benthic assemblages. In: Moraitou-Apostolopoulou, M. and Kiortsis, V. (eds.) *Mediterranean marine ecosystems.* New York: Plenum Press, pp. 19–48.

Bello, G., 2000. How rare is *Histioteuthis bonnellii* (Cephalopoda: Histioteuthidae) in the eastern Mediterranean Sea? *J. Moll. Stud.* 66, 575-577.

Bello, G., 2003. The biogeography of Mediterranean cephalopods. *Biogeographia* 33, 209-226.

Bello, G., 2008. Cephalopoda. *Biol. Mar. Medit.* 15, 318-322.

Benzohra, M., Millot C., 1995. Characteristics and circulation of the surface and intermediate water masses off Algeria. *Deep Sea Res. Pt. I* 28, 1-52.

Berger, W. H., 1989. Appendix. Global maps of ocean productivity. In: Berger, W. H., Smetacek, V. S. and Wefer, G. (eds.) *Productivity of the ocean: present and past.* Chichester: Wiley and Sons, pp. 429-455.

Bethoux, J. P., Gentili, B., Morin, P., Nicolas, E., Pierre, C., Ruiz-Pino, D., 1999. The Mediterranean Sea: a miniature ocean for climatic and environmental studies and a key for the climatic functioning of the North Atlantic. *Prog. Oceanogr.* 44, 131-146.

Bethoux, J. P., Gentili, B., Raunet, J., Tailliez, D., 1990. Warming Trend in the Western Mediterranean Deep-Water. *Nature* 347, 660-662.

Bianchi, C. N., Morri, C., 2000. Marine Biodiversity of the Mediterranean Sea: Situation, Problems and Prospects for Future Research. *Mar. Poll. Bull.* 40, 367-376.

Blake, J. A., 1993. Life history analysis of 5 dominant infaunal polychaete species from the continental slope off north Carolina. *J. Mar. Biol. Assoc. UK* 73(1), 123-141.

Bo, M., 2008. Taxonomy and ecology of antipatharians. Università Politecnica delle Marche, *PhD Thesis*, Universita Politecnica delle Marche, Ancona, Italy, 212 pp.

Bo, M., Bavestrello, G., Canese, S., Giusti, M., Salvati, E., Angiolillo, M., Greco, S., 2009. Characteristics of a black coral meadow in the twilight zone of the central Mediterranean Sea. *Mar. Ecol. Prog. Ser.* 397, 53-61.

Bo, M., Bertolino, M., Bavestrello, G., Canese, S., Giusti, M., Angiolillo, M., Pansini, M., Taviani, M., 2012. Role of deep sponge grounds in the Mediterranean Sea: a case study in southern Italy. *Hydrobiologia*, 687, 163-177.

Bo, M., Bertolino, M., Borghini, M., Castellano, M., Harriague, A. C., Di Camillo, C. G., Gasparini, G., Misic, C., Povero, P., Pusceddu, A., et al., 2011. Characteristics of the mesophotic megabenthic assemblages of the Vercelli seamount (North Tyrrhenian Sea). *Plos One* 6 (2), e16357.

Bouchet, P., Taviani, M., 1992. The Mediterranean deep-sea fauna: pseudopopulations of Atlantic species? *Deep Sea Res. Pt. I* 39, 169-184.

Brooke, S., Young, C. M., 2005. Embryogenesis and larval biology of the ahermatypic scleractinian *Oculina varicosa*. *Mar. Biol.* 146, 665-675.

Brooke, S., Young, C. M., 2009. In situ measurement of survival and growth of *Lophelia pertusa* in the northern Gulf of Mexico. *Mar. Ecol. Prog. Ser.* 397, 153-161.

Bryan, T. L., Metaxas, A., 2007. Predicting suitable habitat for deep-water gorgonian corals on the Atlantic and Pacific Continental Margins of North America. *Mar. Ecol. Prog. Ser.* 330, 113-126.

Buhl-Mortensen, L., Vanreusel, A., Gooday, A. J., Levin, L. A., Priede, I. G., Buhl-Mortensen, P., Gheerardyn, H., King, N. J., Raes, M., 2010. Biological structures as a source of habitat heterogeneity and biodiversity on the deep ocean margins. *Mar. Ecol.* 31, 21-50.

Bulman, C. M., Koslow, J. A., 1992. Diet and food consumption of a deep-sea fish, orange roughy Hoplostethus atlanticus (Pisces: Trachichthydae), off southeastern Australia. *Mar. Ecol. Prog. Ser.* 82, 115-129.

Carbonell, A., Abelló, P., 1998. Distribution characteristics of pandalid shrimps (Decapoda: Caridea: Pandalidae) along the western Mediterranean Sea. *J. Nat. Hist.* 32 (10-11), 1463-1474.

Carbonell, A., Carbonell, M., Demestre, M., Grau, A., Monserrat, S., 1999. The red shrimp *Aristeus antennatus* (Risso, 1816) fishery and biology in the Balearic Islands, Western Mediterranean. *Fish. Res.* 44(1), 1-13.

Carlier, A., Le Guilloux, E., Olu, K., Sarrazin, J., Mastrototaro, F., Taviani, M., Clavier, J., 2009. Trophic relationships in a deep Mediterranean cold-water coral bank (Santa Maria di Leuca, Ionian Sea). *Mar. Ecol. Prog. Ser.* 397, 125-137.

Carpine, C., Grasshoff, M., 1975. Les gorgonaires de la Méditerranée. *Bull. Inst. Océanogr.* Monaco 71, 1–140.

Carrasón, M., Matallanas, J., 1998. Feeding habits of *Alepocephalus rostratus* (Pisces: Alepocephalidae) in the western Mediterranean Sea. *J. Mar. Biol. Ass. UK* 78, 1295-1306.

Carrasón, M., Matallanas, J., 2001. Feeding ecology of the Mediterranean spiderfish *Bathypterois mediterraneus* (Pisces: Chlorophtalmidae) on the western Mediterranean slope. *Fish. Bull.* 99, 266-274.

Carrasón, M., Matallanas, J., 2002. Diets of deep-sea macrourid fish in the western Mediterranean slope. *Mar. Ecol. Progr. Ser.* 234, 215-228.

Carrasón, M., Matallanas, J., Casadevall, M., 1997. Feeding strategies of deep-water morids on the western Mediterranean slope. *Deep-Sea Res. Pt. I* 44, 1685-1699.

Cartes, J. E., 1998. Dynamics of the bathyal Benthic Boundary Layer in the northwestern Mediterranean: depth and temporal variations in macrofaunal-megafaunal communities and their possible connections within deep-sea trophic webs. *Prog. Oceanogr.* 41, 111-139.

Cartes, J. E., Maynou, F., Fanelli, E., Romano, C., Mamouridis, V., Papiol, V., 2009. The distribution of megabenthic, invertebrate epifauna in the Balearic Basin (western Mediterranean) between 400 and 2300 m: Environmental gradients influencing assemblages composition and biomass trends. *J. Sea Res.* 61, 244-257.

Cartes, J. E., Sardà, F., 1992. Abundance and diversity of decapod crustaceans in the deep-Catalan sea (Western Mediterranean). *J. Nat. Hist.* 26(6), 1305-1323.

Cartes, J. E., Company, J. B., Maynou, F., 1994. Deep-water decapod crustacean communities in the Northwestern Mediterranean: influence of submarine canyons and season. *Mar. Biol.* 120, 221-229.

Cerrano, C., Danovaro, R., Gambi, C., Pusceddu, A., Riva, A., Schiaparelli, S., 2010. Gold coral (*Savalia savaglia*) and gorgonian forests enhance benthic biodiversity and ecosystem functioning in the mesophotic zone. *Biodiv. Cons.* 19, 153-167.

Cheng, H., Adkins, J., Edwards, R. L., Boyle, E. A., 2000. U-Th dating of deep-sea corals. *Geoch. Cosmoch. Acta* 64, 2401-2416.

Chilari, A., Thessalou-Legaki, M., Petrakis, G., 2005. Population structure and reproduction of the deep-water shrimp *Plesionika martia* (Decapoda: Pandalidae) from the eastern Ionian Sea. *J. Crustacean Biol.* 25(2), 233-241.

CIESM, 2009. Dynamics of Mediterranean Deep Waters. *CIESM Workshop Monographs* 38, 132 p.

CIHEAM, 2002. *Development and agri-food policies in the Mediterranean region. International Centre for Advanced Mediterranean Agronomic Studies, Annual Report*, Paris.

Clarke, M. R., 1996. The role of cephalopods in the world's oceans: general conclusion and the future. *Philos. T. Roy. Soc. B* 351, 1105-1112.

Clarke, M. R., Merrett, N. R., 1972. The significance of squid, whale and other remains from the stomachs of bottom-living deep-sea fishes. *J. Mar. Biol. Ass. UK* 52, 596-603.

Coll, M., Piroddi, C., Steenbeek, J., Kaschner, K., Lasram, F. B., Aguzzi, J., Ballesteros, E., Bianchi, C. N., Corbera, J., Dailianis, T., et al., 2010. The biodiversity of the Mediterranean Sea: estimates, patterns, and threats. *Plos One* 5(8), e11842.

Collins, M. A., Villanueva, R., 2006. Taxonomy, ecology and behaviour of the cirrate octopods. *Oceanogr. Mar. Biol.* 44, 277-322.

Colloca, F., 2002. Life cycle of the deep-water pandalid shrimp *Plesionika edwardsii* (Decapoda, Caridea) in the central Mediterranean Sea. *J. Crustacean Biol.* 22(4), 775-783.

Colloca, F., Carpentieri, P., Balestri, E., Ardizzone, G. D., 2004. A critical habitat for Mediterranean fish resources: shelf-break areas with *Leptometra phalangium* (Echinodermata: Crinoidea). *Mar. Biol.* 145, 1129-1142.

Company, J. B., 1995. Comparative study of the biological trends of crustacean decapod on the Catalan Sea (Mediterranean contiental margin). *PhD dissertation*, University of Barcelona.

Company, J. B., Sardà, F., 1997. Reproductive patterns and population characteristics in five deep-water pandalid shrimps in the Western

Mediterranean along a depth gradient (150-1100 m). *Mar. Ecol. Prog. Ser.* 148(1-3), 49-58.

Company, J. B., Sardà, F., Puig, P., Cartes, J. E., Palanqués, A., 2003. Duration and timing of reproduction in decapod crustaceans of the NW Mediterranean continental margin: is there a general pattern? *Mar. Ecol. Prog. Ser.* 261, 201-216.

Company, J. B., Maiorano, P., Tselepides, A., Politou, C.-Y., Plaity, W., Rotllant, G., Sardà, F., 2004. Deep-sea decapod crustaceans in the western and central Mediterranean Sea: preliminary aspects of species distribution, biomass and population structure. *Sci. Mar.* 68 (Suppl. 3), 73-86.

Crabtree, R. E., Carter, H. J., Musick, J. A., 1991. The comparative feeding ecology of temperature and tropical deep-sea fishes from the western North Atlantic. *Deep-Sea Res.* 38, 1277-1298.

Cuccu, D., Mereu, M., Follesa, M., Deiana, A., Cau, A., 2011. *Bathypolypus sponsalis* (Cephalopoda: Octopoda) from the central western Mediterranean Sea. *J. Mar. Biol. Assoc. UK* 91, 549-553.

D'Onghia, G., Tursi, A., Maiorano, P., Matarrese, A., Panza, M., 1998a. Demersal fish assemblages from the bathyal grounds of the Ionian Sea (middle-eastern Mediterranean). *Ital. J. Zool.* 65, 287-292.

D'Onghia, G., Tursi, A., Marano, C. A., Basanisi, M., 1998b. Life history traits of *Hoplostethus mediterraneus* (Pisces: Beryciformes) from the North-western Ionian Sea (Mediterranean Sea). *J. Mar. Biol. Ass. UK* 78, 321-339.

D'Onghia, G., Basanisi, M., Tursi, A., 2000. Population structure, age and growth of macrourid fish from the upper slope of the Eastern-Central Mediterranean. *J. Fish Biol.* 56, 1217-1238.

D'Onghia, G., Politou, C. Y., Bozzano, A., Lloris, D., Rotllant, G., Sion, L., Mastrototaro, F., 2004. Deep-water fish assemblages in the Mediterranean Sea. *Sci. Mar.* 68, 87-99.

D'Onghia, G., Politou, C.-Y., Bozzano, A., Lloris, D., Rotllant, G., Sion, L., Mastrototaro, F., 2004a. Deepwater fish assemblages in the Mediterranean Sea. *Sci. Mar.* 68 (Suppl. 3), 87-99.

D'Onghia, G., Lloris, D., Sion, L., Capezzuto, F., Labropoulou, M., 2004b. Observations on the distribution, population structure and biology of *Bathypterois mediterraneus* Bauchot, 1962 in three areas of the Mediterranean Sea. *Sci. Mar.* 68 (Suppl. 3), 167-170.

D'Onghia, G., Lloris, D., Politou, C.-Y., Sion, L., Dokos, J., 2004c. New records of deep-water teleost fishes in the Balearic Sea and Ionian Sea (Mediterranean Sea). *Sci. Mar.* 68 (Suppl. 3), 171-183.

D'Onghia, G., Tursi, A., Basanisi, M., 1996. Reproduction of macrourids in the upper slope of the north-western Ionian Sea. *J. Fish Biol.* 49, 311-317.

Dahl, M., Pereyra, R., Lundälv, T., André, C., 2012. Fine-scale spatial genetic structure and clonal distribution of the cold-water coral *Lophelia pertusa*. *Coral Reefs,* 1-14.

Danovaro, R., Company, J., Corinaldesi, C., D'Onghia, G., Galil, B., Gambi, C., Gooday, A. J., Lampadariou, N., Luna, G. M., Morigi, C., Olu, K., Polymenakou, P., Ramirez-Llodra, E., Sabbatini, A., Sarda, F., Sibuet, M., Tselepides, A., 2010. Deep-sea biodiversity in the Mediterranean Sea: the known, the unknown, and the unknowable. *Plos One* 5, e11832.

Danovaro, R., Della Croce, N., Eleftheriou, A., Fabiano, M., Papadopoulou, N., Smith, C., Tselepides, A., 1995. Meiofauna of the deep Eastern Mediterranean Sea: distribution and abundance in relation to bacterial biomass, organic matter composition and other environmental factors. *Prog. Oceanogr.* 36, 329-341.

Danovaro, R., Dell'Anno, A., Fabiano, M., Pusceddu, A., Tselepides, A., 2001. Deep-sea ecosystem response to climate changes: the eastern Mediterranean case study. *Trends Ecol. Evol.* 16, 505-510.

Danovaro, R., Dell'Anno, A., Pusceddu, A., Fangliang, H., 2004. Biodiversity response to climate change in a warm deep sea. *Ecol. Lett.* 7, 821-828.

Danovaro, R., Dinet, A., Duineveld, G., Tselepides, A., 1999. Benthic response to particulate fluxes in different trophic environments: a comparison between the Gulf of Lions - Catalan Sea (western-Mediterranean) and the Cretan Sea (eastern-Mediterranean). *Prog. Oceanogr.* 44, 287-312.

Danovaro, R., Tselepides, A., Otegui, A., Della Croce, N., 2000. Dynamics of meiofaunal assemblages on the continental shelf and deep-sea sediments of the Cretan Sea (NE Mediterranean): relationships with seasonal changes in food supply. *Prog. Oceanogr.* 46, 367-400.

Davies, A. J., Guinotte, J. M., 2011. Global habitat suitability for framework-forming Cold-Water Corals. *Plos One* 6 (4), e18483.

De Madron, X. D., Zervakis, V., Theocharis, A., Georgopoulos, D., 2005. Comments on "Cascades of dense water around the world ocean". *Prog. Oceanogr.* 64, 83-90.

Delibrias, G., Taviani, M., 1985. Dating the death of mediterranean deep-sea scleractinian corals. *Mar. Geol.* 62, 175-180.

Desbruyères, D., Guille, A., Ramos, J., 1972-73. Bionomie benthique du plateau continental de la côte catalane espagnole. *Vie Milieu* 23(2), 335-363.

Dodds, L. A., Black, K. D., Orr, H., Roberts, J. M., 2009. Lipid biomarkers reveal geographical differences in food supply to the cold-water coral *Lophelia pertusa* (Scleractinia). *Mar. Ecol. Progr. Ser.* 397, 113-124.

Douville, E., Salle, E., Frank, N., Eisele, M., Pons-Branchu, E., Ayrault, S., 2010. Rapid and accurate U-Th dating of ancient carbonates using inductively coupled plasma-quadrupole mass spectrometry. *Chem. Geol.* 272, 1-11.

Duineveld, G. C. A., Lavaleye, M. S. S., Bergman, M. J. N., de Stigter, H., Mienis, F., 2007. Trophic structure of a cold-water coral mound community (Rockall Bank, NE Atlantic) in relation to the near-bottom particle supply and current regime. *Bull. Mar. Sci.* 81, 449-467.

Dullo, W. C., Flogel, S., Ruggeberg, A., 2008. Cold-water coral growth in relation to the hydrography of the Celtic and Nordic European continental margin. *Mar. Ecol. Prog. Ser.* 371, 165-176.

Eltgroth, S. F., Adkins, J. F., Robinson, L. F., Southon, J., Kashgarian, M., 2006. A deep-sea coral record of North Atlantic radiocarbon through the Younger Dryas: Evidence for intermediate water/deepwater reorganization. *Paleoceanography* 21 (4), PA4207.

Estrada, M., 1996. Primary production in the northwestern Mediterranean. *Sci. Mar.* 60, 55-64.

Fanelli, E., Cartes, J. E., Papiol, V., 2012. Assemblage structure and trophic ecology of deep-sea demersal cephalopods in the Balearic basin (NW Mediterranean). *Mar. Freshwater Res.* 63, 264-274.

Fanelli, E., Colloca, F., Ardizzone, G., 2007. Decapod crustacean assemblages off the West coast of central Italy (western Mediterranean). *Sci. Mar.* 71 (1), 19-28.

Fanelli, E., Rey, J., Torres, P., de Sola, L. G., 2009. Feeding habits of blackmouth catshark *Galeus melastomus* Rafinesque, 1810 and velvet belly lantern shark *Etmopterus spinax* (Linnaeus, 1758) in the western Mediterranean. *J. Appl. Ichthyol.* 25, 83-93.

Fernandez-Arcaya, U., Recasens, L., Murua, H., Ramirez-Llodra, E., Rotllant, G., Company, J. B., 2012. Population structure and reproductive patterns of the NW Mediterranean deep-sea macrourid *Trachyrincus scabrus* (Rafinesque, 1810). *Mar. Biol.* 159(9), 1885-1896.

Fishelson, L., Galil, B. S., 2001. Gonad structure and reproductive cycle in the deep-sea hermaphrodite tripodfish, *Bathypterois mediterraneus* (Chlorophthalmidae, Teleostei). *Copeia* 2, 556-560.

Follesa, M. C., Cabiddu, S., Gastoni, A., Cau, A., 2007. On the reproductive biology of the deep-sea lobster, *Polycheles typhlops* (Decapoda, Palinura,

Polychelidae), from the central-western Mediterranean. *Crustaceana* 80 (7), 839-846.

Follesa, M. C., Porcu, C., Gastoni, A., Mulas, A., Sabatini, A., Cau, A., 2009. Community structure of bathyal decapods crustaceans off South-Eastern Sardinian deep-waters (Central-Western Mediterranean). *Mar. Ecol.-Evol. Persp.* 30, 188-199.

Frank, N., Lutringer, A., Paterne, M., Blamart, D., Henriet, J. P., van Rooij, D., van Weering, T. C. E., 2005. Deep-water corals of the northeastern Atlantic margin: carbonate mound evolution and upper intermediate water ventilation during the Holocene. In: Freiwald, et al. (eds.) Cold-water corals and ecosystems. *Erlangen Earth Conference Series* pp: 113-133.

Fredj, G., Bellan-Santini, D., Meinardi, M., 1992. État des connaissances sur la faune marine méditeranéenne. In: Bellan, D. (Ed.) Spéciation et biogéographie en mer Méditerranée. *Bull. Inst. Océanogr. (Monaco)* Spécial 9, 133-145.

Fredj, G., Laubier, L., 1985. The deep Mediterranean benthos. In: Moraitou-Apostolopoulou, M. and Kiortis, V. (eds.) *Mediterranean marine ecosystems.* New York: Plenum Press, pp. 109-145.

Fredj, G., Maurin, C., 1987. Les poisons dans la banque de données MEDIFAUNE. Application à l'étude des caractéristiques de la faune ichtyologique méditerranéenne. *Cybium* 11(3), 31-139.

Freiwald, A., Beuck, L., Rueggeberg, A., Taviani, M., Hebbeln, D., 2009. The white coral community in the Central Mediterranean Sea Revealed by ROV Surveys. *Oceanography* 22, 58-74.

Gage, J., 2003. Food inputs, utilisation, carbon flow and energetics. In: Tyler, P. A. (Ed.) *Ecosystems of the world: The deep ocean.* Elsevier, Amsterdam, 313-426.

Gage, J. D., Tyler, P. A., 1991. *Deep-sea biology. A Natural History of Organisms at the Deep-Sea Floor.* Cambridge University Press, 504 pp.

Galgani, F., Souplet, A., Cadiou, Y., 1996. Accumulation of debris on the deep sea floor off the French Mediterranean coast. *Mar. Ecol. Prog. Ser.* 142, 225-234.

Galil, B., Goren, M., 1994. The Deep Sea Levantine Fauna. New Records and Rare Occurrences. *Senckenbergiana Maritima* 25(1/3), 41-52.

Galil, B. S., Golik, A., Turkay, M., 1995. Litter at the Bottom of the Sea - A Sea-Bed Survey in the Eastern Mediterranean. *Mar. Poll. Bull.* 30, 22-24.

García Lafuente, J., Sanchez Roman, A., Diaz del Rio, G., Sannino, G. Sanchez Garrido, J. C., 2007. Recent observations of seasonal variability of the Mediterranean outflow in the Strait of Gibraltar. *J. Geophys. Res.*, 112, C10005, doi: 10.1029/2006JC003992.

García-Rodríguez, M., Esteban, A., Pérez Gil, J. L., 2000. Considerations on the biology of *Plesionika edwardsi* (Brandt, 1851) (Decapoda, Caridea, Pandalidae) from experimental trap catches in the Spanish western Mediterranean Sea. *Sci. Mar.* 64(4), 369-379.

Gass, S. E., Roberts, J. M., 2011. Growth and branching patterns of *Lophelia pertusa* (Scleractinia) from the North Sea. *J. Mar. Biol. Ass. UK* 91, 831-835.

Gerin, R., Poulain, P. M., Taupier-Letage, I., Millot, C., Ben Ismail, S., Sammari, C., 2009. Surface circulation in the Eastern Mediterranean using drifters (2005–2007). *Ocean Sci.* 5, 559–574.

GFCM-FAO, 2011. *General fisheries commission for the Medietrranean. Fifth session of the compliance committee (CoC). Thirty-fifth session of the commission.* Rome, Italy, 9-14 May 2011. *Compendium of GFCM decisions.* REC.CM-GFCM/30/2006/3. 206 pp.

Giangrande, A., Geraci, S., Belmonte, G., 1994. Life-cycle and life-history diversity in marine invertebrates and the implications in community dynamics. *Oceanog. Mar. Biol. Ann. Rev.* 32, 305-333.

Gili, J. M., Madurell, T., Requena, S., Orejas, C., Gori, A., Purroy, A., Domínguez, C., Lo Iacono, C., Isla, E., Lozoya, J. P., Carboneras, C., Grinyó, J., 2011. *Caracterización física y ecológica del área marina del Cap de Creus. Informe final área LIFE+INDEMARES (LIFE07/NAT/E/ 000732).* Instituto de Ciencias del Mar/CSIC (Barcelona). Coordinación, Fundación Biodiversidad: Madrid, 272 pp.

González, M., Sánchez, P., 2002. Cephalopod assemblages caught by trawling along the Iberian Peninsula Mediterranean coast. *Sci. Mar.* 66, 199-208.

Goren, M., Galil, B. S., 1997. New records of deep-sea fishes from the Levant basin and a note on the deep-sea fishes of the Mediterranean. *Israel J. Zool.* 43, 197-203.

Gori, C., Orejas, C., Madurell, T., Bramanti, L., Martins, M., Quintanilla, E., Marti-Puig, P., Lo Iacono, C., Puig, P., Requena, S., Greenacre, M., Gili, J. M., 2013. Bathymetrical distribution and size structure of cold-water coral populations in the Cap de Creus and Lacaze-Duthiers canyons (northwestern Mediterranean). *Biogeosciences* 10, 2049-2060.

Guijarro, B., Massutí, E., 2006. Selectivity of diamond- and square-mesh codends in the deepwater crustacean trawl fishery off the Balearic Islands (western Mediterranean). *ICES J. Mar. Sci.* 63, 52-67.

Guijarro, B., Massutí, E., Moranta, J., Díaz, P., 2008. Population dynamics of the red-shrimp Aristeus antennatus in the Balearic Islands (western Mediterranean): spatio-temporal differences and influence of environmental factors. *J. Marine Syst.* 71, 385-402.

Guijarro, B., Massutí, E., Moranta, J., Cartes, J. E., 2009. Short spatio-temporal variations in the population dynamics and biology of the deep-water rose shrimp *Parapenaeus longirostris* (Decapoda: Crustacea) in the western Mediterranean. *Sci. Mar.* 73(1), 183-197.

Guijarro, B., Tserpes, G., Moranta, J., Massutí, E., 2011. Assessment of the deep water trawl fishery off the Balearic Islands (western Mediterranean): from single to multi-species approach. *Hydrobiologia* 670, 67-85.

Guijarro, B., 2012. Population dynamics and assessment of exploited deep water decapods off Balearic Islands (western Mediterranean): from single to multi-species approach. *PhD dissertation*, University of the Balearic Islands, Spain.

Guijarro, B., Fanelli, E., Moranta, J., Cartes, J. E., Massutí, E., 2012. Small-scale differences in the distribution and population dynamics of pandalid shrimps in the western Mediterranean in relation to environmental factors. *Fish. Res.* 119-120, 33-47.

Guillen, J., Maynou, F., Floros, C., Sampson, D., Conides, A., Kapiris, K., 2012. A bio-economic evaluation of the potential for establishing a commercial fishery on two newly developed stocks: The Ionian red shrimp fishery. *Sci. Mar.* 76(3), 597-605.

Haedrich, R. L., Henderson, N. R., 1974. Pelagic food of *Coryphaenoides rupestris*. *Deep-Sea Res.* 21, 739-744.

Haedrich, R. L., Merrett, N. R., 1992. Production/biomass ratios, size frequencies, and biomass spectra in deep-sea demersal fishes. In: G. T. Rowe G. T. and Pariente, V. (eds.) *Deep-Sea Food Chains and the Global Carbon Cycle*. Netherlands: Kluwer Academic Publishers, pp. 157-182.

Hamad, N., Millot, C., Taupier-Letage, I., 2006. The surface circulation in the eastern basin of the Mediterranean Sea. *Sci. Mar.* 70(3), 457–503.

Harmelin, J. G., d'Hondt, J. L., 1993. Transfers of bryozoan species between the Atlantic Ocean and the Mediterranean Sea via the Strait of Gibraltar. *Oceanol. Acta* 16 (1), 63-72.

Henry, L. A., Davies, A. J., Roberts, J. M., 2010. Beta diversity of cold-water coral reef communities off western Scotland. *Coral Reefs* 29, 427-436.

Houlbreque, F., McCulloch, M., Roark, B., Guilderson, T., Meibom, A., Kimball, J., Mortimer, G., Cuif, J. P., Dunbar, R., 2010. Uranium-series dating and growth characteristics of the deep-sea Scleractinian coral:

Enallopsammia rostrata from the Equatorial Pacific. *Geoch. Cosmoch. Acta* 74, 2380-2395.

Jonsson, L. G., Nilsson, P. G., Floruta, F., Lundalv, T., 2004. Distributional patterns of macro- and megafauna associated with a reef of the cold-water coral *Lophelia pertusa* on the Swedish west coast. *Mar. Ecol. Prog. Ser.* 284, 163-171.

Kallianiotis, A., Sophronidis, K., Vidoris, P., Tselepides, A., 2000. Demersal fish and megafaunal assemblages on the Cretan continental shelf and slope (NE Mediterranean): seasonal variation in species density, biomass and diversity. *Progr. Oceanogr.* 46, 429-455.

Kapiris, K., Kavvadas, S., 2009. Morphometric study of the red shrimp *Aristeus antennatus* in the Eastern Mediterranean. *Aquat. Ecol.* 43(4), 1061-1071.

Katz, E. J., 1972. The Levantine Intermediate water between the Strait of Sicily and the Strait of Gibraltar. *Deep-Sea Res.* 19, 507-520.

Keller, N. B., Os'kina, N. S., 2008. Habitat temperature ranges of azooxantellate scleractinian corals in the World Ocean. *Oceanology* 48, 77-84.

Kiriakoulakis, K., Fisher, E., Wolff, G. A., Freiwald, A., Grehan, A., Roberts, J. M., 2005. Lipids and nitrogen isotopes of two deep-water corals from the North-East Atlantic: initial results and implications for their nutrition. In: Freiwald, et al. (eds.) Cold-water corals and ecosystems. *Erlangen Earth Conference Series* pp: 715-729.

Koukouras, A., Kitsos, M. S., Tselepides, A., 2000. The genera *Chaceon* Manning and Holthius and *Geryon* Kroyer (Decapoda, Geryonidae) in the eastern Mediterranean. *Crustaceana* 73, 107-113.

Lapian, H. F. N., Barucca, M., Bavestrello, G., Biscotti, M. A., Bo, M., Canapa, A., Tazioli, S., Olmo, E., 2007. A systematic study of some Black Corals species (Antipatharia, Hexacorallia) based on rDNA internal transcribed spacers sequences. *Mar. Biol.* 151(2), 785-792.

Larsson, A. I., Purser, A., 2011. Sedimentation on the cold-water coral Lophelia pertusa: Cleaning efficiency from natural sediments and drill cuttings. *Mar. Pollut. Bull.* 62, 1159-1168.

Lartaud, F., Pareige, S., de Rafelis, M., Feuillassier, L., Bideau, M., Peru, E., Romans, P., Alcala, F., Le Bris, N., 2013. A new approach for assessing cold-water coral growth in situ using fluorescent calcein staining. *Aquat. Living Res.* 26, 187-196.

Lascaratos, A., Roether, W., Nittis, K., and Klein, B. (1999). Recent changes in deep water formation and spreading in the eastern Mediterranean Sea: a review. *Prog. Oceanogr.* 44(1-3), 5-36.

Laubier, L., Ewig, C. C., 1993. La faune benthique profonde de la Méditerranée. In: Istituto di Scienze Ambientali Marine (ed.) *Symposium on Mediterranean Seas 2000*. Santa Margherita Ligure (Italy), 23–27 September 1991, pp. 397–424.

Lefkaditou, E., Mytilineou, Ch., Maiorano, P., D´Onghia, G., 2003. Cephalopod species captured by deep-water exploratory trawling in the Eastern Ionian Sea. *J. Northw. Atl. Fish Sci.* 31, 431-440.

Lejeusne, C., Chevaldonné, P., Pergent-Martini, C., Boudouresque, C. F., Pérez, T., 2010. Climate change effects on a miniature ocean: the highly diverse, highly impacted Mediterranean Sea. *Trends Ecol. Evol.* 25, 250-260.

Levin, L. A., Etter, R. J., Rex, M. A., Gooday, A. J., Smith, C. R., Pineda, J., Stuart, C. T., Hessler, R. R., Pawson, D., 2001. Environmental influences on regional deep-sea species diversity. *Annu. Rev. Ecol. Syst.* 32, 51-93.

Lo Iacono, C., Gracia, E., Diez, S., Bozzano, G., Moreno, X., Danobeitia, J., Alonso, B., 2008. Seafloor characterization and backscatter variability of the Almeria Margin (Alboran Sea, SW Mediterranean) based on high-resolution acoustic data. *Mar. Geol.* 250, 1-18.

Lo Iacono, C., Orejas, C., Gori, A., Gili, J. M., Requena, S., Puig, P., Ribó, M., 2012. Habitats of the Cap de Creus continental shelf and Cap de Creus canyon, north-western Mediterranean. In: *Seafloor geomorphology as benthic habitat: GeoHab Atlas of seafloor geomorphic features and benthic habitats*, (eds. P. T. Harris and E. K. Baker), pp. 457-470. London: Elsevier.

López Correa, M., Montagna, P., Joseph, N., Ruggeberg, A., Fietzke, J., Floegel, S., Dorschel, B., Goldstein, S. L., Wheeler, A., Freiwald, A., 2012. Preboreal onset of cold-water coral growth beyond the Arctic Circle revealed by coupled radiocarbon and U-series dating and neodymium isotopes. *Quaternary Sci. Rev.* 34, 24-43.

López-Jurado, J. L., Gonzalez-Pola, C., Velez-Belchi, P., 2005. Observation of an abrupt disruption of the long-term warming trend at the Balearic Sea, western Mediterranean Sea, in summer 2005. *Geophys. Res. Lett.* 32, L 24606, doi:10.1029/2005GL024430.

Lumsden, S. E., Hourigan, T. F., Bruckner, A. W., Dorr, G., 2007. *The State of Deep Coral Ecosystems of the United States: NOAA Technical Memorandum CRCP-3.* Silver Spring MD.

Madurell, T., Cartes, J. E., Labropoulou, M., 2004. Changes in the structure of fish assemblages in a bathyal site of the Ionian Sea (eastern Mediterranean). *Fish. Res.* 66, 245-260.

Madurell, T., Cartes, J. E., 2005. Temporal changes in feeding habits and daily rations of *Hoplostethus mediterraneus*, Cuvier, 1829 in the bathyal Ionian Sea (Eastern Mediterranean). *Mar. Biol.* 146, 951-962.

Madurell, T., Cartes, J. E., 2006. Trophic relationships and food consumption of slope dwelling macrourids from the bathyal Ionian Sea (eastern Mediterranean). *Mar. Biol.* 148, 1325-1338.

Madurell, T., Gori, A., Requena, S., Gili, J. M., Orejas, C., Quintanilla, E., Ferrari, B., Watremez, P., Romans, P., Fiala-Médioni, A., 2010. *Cold water coral assemblages in Lacaze-Duthiers Canyon (Western Mediterranean), oral presentation, HERMIONE annual meeting*, Malta.

Maier, C., Hegeman, J., Weinbauer, M. G., Gattuso, J. P., 2009. Calcification of the cold-water coral *Lophelia pertusa* under ambient and reduced pH. *Biogeosciences* 6, 1671-1680.

Maier, C., Watremez, P., Taviani, M., Weinbauer, M. G., Gattuso, J. P., 2011. Calcification rates and the effect of ocean acidification on Mediterranean cold-water corals. *Proc. R. Soc. B* doi:10.1098/rspb.2011.1763.

Maiorano, P., D'Onghia, G., Capezzuto, F., Sion, L., 2002. Life-history traits of *Plesionika martia* (Decapoda: Caridea) from the eastern-central Mediterranean sea. *Mar. Biol.* 141 (3), 527-539.

Malanotte-Rizzoli, P., Manca, B., Ribera D'Alcala, M., Theocharis, A., Bergamasco, A., Bregant, D., Budillon, G., Civitarese, G., Georgopoulos, D., Michelato, A., Sansone, E., Scarazzato, P., Souvermezoglou, E., 1997. A synthesis of the Ionian Sea hydrography, circulation and water mass pathway during POEM-Phase I. *Prog. Oceanogr.* 39, 153–204.

Maldonado, A., 1985. Evolution of the Mediterranean basins and a detailed reconstruction of the Cenozoic paleoceanography. In: Margalef, R. (Ed.) *Western Mediterranean*. Pergamon Press Ltd., Oxford, 17-59.

Manca, B., Burca, M., Giorgetti, A., Coatanoan, C., Garcia, M. J., Iona, A., 2004. Physical and biochemical averaged vertical profiles in the Mediterranean regions: an important tool to trace the climatology of water masses and to validate incoming data from operational oceanography. *J. Marine Syst.* 48, 83-116.

Mangini, A., Lomitschka, M., Eichstadter, R., Frank, N., Vogler, S., Bonani, G., Hajdas, I., Patzold, J., 1998. Coral provides way to age deep water. *Nature* 392, 347-348.

Mangold-Wirz, K., 1973. Distribution géographique de Pteroctopus tetracirrhus (Delle Chiaje). Contribution au problème de la taille des oeufs chez les Octopodidae. *Rapp. Comm. Int. Mer Médit.* 21, 785-787.

Mangold-Wirz, K., Boletzky, S. V., Mesnil, B., 1976. Biologie de reproduction et distribution d´Octopus salutii Verany (Cephalopoda, Octopoda). *Rapp. Comm. Int. Mer Médit.* 23, 87-93.

Margalef, R., 1985. Introduction to the Mediterranean. In: Margalef, R. (Ed.) *Western Mediterranean*. Pergamon Press Inc., Oxford, 1-16.

Marin, P., Pastor, X., Aguilar, R., Garcia, S., Pardo, E., Ubero, J., 2011. Montañas submarinas de las Islas Baleares: Canal de Mallorca 2011. Propuesta de protección para Ausías March, Emile Baudot y Ses Olives. Oceana. 60 pp.

Masó M., La Violette P. E., Tintoré J., 1990. Coastal flow modification by submarine canyons along the NE Spanish coast. *Sci. Mar.* 54(4), 343-348.

Massutí, E., Morales-Nin, B., Stefanescu, C., 1995. Distribution and biology of five grenadier fish (Pisces: Macrouridae) from the upper and middle slope of the northwestern Mediterranean. *Deep-Sea Res.* 42, 307-330.

Massutí, E., Morales-Nin, B., Lloris, D., 1996. Bathymetric distribution and recruitment patterns of *Phycis blennoides* (Pisces: Gadidae) from the slope of the northwestern Mediterranean. *Sci. Mar.* 60, 481-488.

Massutí, E., Gordon, J. D. M., Moranta, J., Swan, S. C., Stefanescu, C., Merrett, N. R., 2004. Mediterranean and Atlantic deep-sea fish assemblages: differences in biomass composition and size-related structure. *Sci. Mar.* 68 (Suppl. 3), 101-115.

Mastrotaro, F., D'Onghia, G., Corriero, G., Matarrese, A., Maiorano, P., Panetta, P., Gherardi, M., Longo, C., Rosso, A., Sciuto, F., et al., 2010. Biodiversity of the white coral bank off Cape Santa Maria di Leuca (Mediterranean Sea): An update. *Deep-Sea Res. Pt. II* 57, 412-430.

Maurin, C. L., 1968. *Ecologie ichthyologique des fonds chaliitables atlantiques (de la baie ibéro-marocaine A la Mauiritanie) et de la Méditerranée occidentale*, 1e These Université Nancy.

Maynou, F., Conan, G. Y., Cartes, J. E., Company, J. B., Sardà, F., 1996. Spatial structure and seasonality of decapod crustacean populations on the northwestern Mediterranean slope. *Limnol. Oceanogr.* 41(1), 113-125.

Maynou, F., Cartes, J. E., 2000. Community structure of bathyal decapod crustaceans off south-west Balearic Islands (western Mediterranean): seasonality and regional patterns in zonation. *J. Mar. Biol. Ass. UK* 80, 789-798.

McCulloch, M. T., Taviani, M., Montagna, P., Mortimer, G., Remia, A., 2006. Proliferation and demise of Mediterranean deep-sea corals. *Geoch. Cosmoch. Acta* 70, A407-A407.

McKenzie, J. A., 1999. From desert to deluge in the Mediterranean. *Nature* 400, 613-614.

MEDOC Group, 1970. Observation of formation of deep water in the Mediterranean Sea. *Nature* 277, 1037–1040.

Merrett, N. R., Haedrich, R. L., 1997. *Deep-sea demersal fish and fisheries*. London: Chapman and Hall, 282 pp.

Mienis, H. K., 2003. Mariene mollusken uit het oostelijk deel van de Middellandse Zee-15. Invasie van rugschilden van Sepia pharaonis langs de kust van Israël. *Spirula* 335, 127-129.

Millot, C., 1999. Circulation in the western Mediterranean Sea. *J. Marine Syst.* 20, 423-442.

Miquel, J. C., Fowler, S. W., La Rosa, J., Buat-Menard, P., 1994. Dynamics of the downward flux of particles and carbonin the open northwestern Mediterranean Sea. *Deep-Sea Res.* 41, 243-261.

Montagna, P., McCulloch, M., Taviani, M., Remia, A., Rouse, G., 2005. High-resolution trace and minor element compositions in deep-water scleractinian corals (*Desmophyllum dianthus*) from the Mediterranean Sea and the Great Australian Bight. In: Freiwald, et al. (eds.) Cold-water corals and ecosystems. *Erlangen Earth Conference Series* pp: 1109-1126.

Montagna, P., Silenzi, S., Devoti, S., Mazzoli, C., McCulloch, M., Scicchitano, G., Taviani, M., 2008. Climate reconstructions and monitoring in the Mediterranean Sea: a review on some recently discovered high-resolution marine archives. *Rendiconti Lincei* 19, 121-140.

Montero-Serrano, J.-C., Frank, N., Colin, C., Wienberg, C., Eisele, M., 2011. The climate influence on the mid-depth Northeast Atlantic gyres viewed by cold-water corals. *Geoph. Res. Lett.* 38(19) doi: 10.1029/2011GL0487 33.

Morales-Nin, B., Massutí, E., Stefanescu, C., 1996a. Bathymetric distribution and growth patterns of *Bathypterois mediterraneus* from the north-western Mediterranean Sea. *J. Fish Biol.* 49 (supplement A), 276-288.

Morales-Nin, B., Massutí, E., Stefanescu, C., 1996b. Distribution and biology of *Alepocephalus rostratus* from the Mediterranean Sea. *J. Fish Biol.* 48, 1097-1112.

Moranta, J., Stefanescu, C., Massutí, E., Morales-Nin, B., Lloris, D., 1998. Fish community structure and depth-related trends on the continental slope of the Balearic Islands (Algerian basin, western Mediterranean). *Mar. Ecol. Prog. Ser.* 171, 247-259.

Moranta, J., Massuti, E., Morales-Nin, B., 2000. Fish catch composition of the deep-sea decapod crustacean fisheries in the Balearic Islands (western Mediterranean). *Fish. Res.* 45, 253-264.

Moranta, J., Quetglas, A., Massuti, E., Guijarro, B., Ordines, F., Valls, M., 2008. Research trends on demersal fisheries oceanography in the Mediterranean. In: Mertens, L. P. (Ed.) *Biological Oceanography Research Trends.* Nova Science Publishers, Inc., New York, 9-65.

Movilla, J., Gori, A., Calvo, E., Orejas, C., López-Sanz, A., Domínguez-Carrió, C., Grinyó, J., Pelejero, C., 2014a. Resistance of two Mediterranean Cold-Water Coral species to low-pH conditions. *Water* 6, 59-67.

Movilla, J., Orejas, C., Calvo, E., Gori, A., López-Sanz, A., Grinyó, J., Dominguez-Carrió, C., Pelejero, C., 2014b. Differential response of two Mediterranean cold-water coral species to ocean acidification. *Coral Reefs* DOI 10.1007/s00338-014-1159-9.

Mytilineou, Ch., Maiorano, P., Kavadas, S., D'Onghia, G., Kapiris, K., Capezzuto, F., 2001. Size structure comparison in some demersal species between two areas of different fishing impact in the deep waters of the eastern-central Mediterranean (Ionian Sea). *J. Northw. Atl. Fish. Sci.* NAFO SCR Doc. 01/125.

Mytilineou, Ch., Kavadas, S., Politou C.-Y., Kapiris, K., Tursi, A., Maiorano, P., 2006. Catch composition on red shrimps' (*Aristaeomorpha foliacea* and *Aristeus antennatus*) grounds in the Eastern Ionian Sea. *Hydrobiologia* 557, 155-160.

Naumann, M. S., Orejas, C., Wild, C., Ferrier-Pagès, C., 2011. First evidence for zooplankton feeding sustaining key physiological processes in a Scleractinian cold-water coral. *J. Exp. Biol.* 214, 3570-3576.

Naumann, M. S., Orejas, C., Ferrier-Pagès, C. 2014. Species-specific response in key physiological processes by the cold-water corals *Lophelia pertusa* and *Madrepora oculata* within their natural temperature range. *Deep Sea Res. Pt. II* 99, 36-41.

Naumann, M. S., Orejas, C., Ferrier-Pagès, C., 2013. High thermal tolerance of two Mediterranean cold-water coral species maintained in aquaria. *Coral Reefs* doi 10.1007/s00338-013-1011-7.

Norman, M., 2003. *Cephalopods, A world guide.* ConchBooks, Hackenheim.

Nouar, A., Maurin, C., 2001. Nature of and typical populations on the characteristic facies of substratum of *Parapenaeus longirostris* (Lucas, 1846) along the Algerian coast. *Crustaceana* 74: 129-135.

Olivar, M. P., Bernal, A., Molí, B., Peña, M., Balbín, R., Castellón, A., Miquel, J., Massutí, E., 2012. Vertical distribution, diversity and assemblages of mesopelagic fishes in the western Mediterranean. *Deep-Sea Res. Pt. I* 62, 53-69.

Ordines, F., Massutí, E., 2009. Relationships between macro-epibenthic communities and fish on the shelf grounds of the western Mediterranean. *Aquat. Conserv. Mar. Freshw. Ecosyst.* 19, 370-383.

Orejas, C., López-González, P. J., Gili, J. M., Teixido, N., Gutt, J., Arntz, W. E., 2002. Distribution and reproductive ecology of the Antarctic octocoral *Ainigmaptilon antarcticum* in the Weddell Sea. *Mar. Ecol. Prog. Ser.* 231, 101-114.

Orejas, C., Gili, J. M., López-González, P. J., Hasemann, C., Arntz, W. E., 2007. Reproduction patterns of four Antarctic octocorals in the Weddell Sea: an inter-specific, shape, and latitudinal comparison. *Mar. Biol.* 150, 551-563.

Orejas, C., Gori, A., Gili, J. M., 2008. Growth rates of live *Lophelia pertusa* and *Madrepora oculata* from the Mediterranean Sea maintained in aquaria. *Coral Reefs* 27, 255-255.

Orejas, C., Gori, A., Lo Iacono, C., Puig, P., Gili, J. M., Dale, M. R. T., 2009. Cold-water corals in the Cap de Creus canyon, northwestern Mediterranean: spatial distribution, density and anthropogenic impact. *Mar. Ecol. Prog. Ser.* 397, 37-51.

Orejas, C., Ferrier-Pagès, C., Reynaud, S., Gori, A., Beraud, E., Tsounis, G., Allemand, D., Gili, J. M., 2011a. Long-term growth rates of four Mediterranean cold-water coral species maintained in aquaria. *Mar. Ecol. Prog. Ser.* 429, 57-65.

Orejas, C., Ferrier-Pagès, C., Reynaud, S., Tsounis, G., Allemand, D., Gili, J. M., 2011b. Experimental comparison of skeletal growth rates in the cold-water coral *Madrepora oculata* Linnaeus, 1758 and three tropical scleractinian corals. *J. Exp. Mar. Biol. Ecol.* 405, 1-5.

Orsi-Relini, L., Belluscio, A., Ardizzone, G. D., 2004. Tracking the Indopacific octopus *Tremoctopus gracilis* in the Mediterranean. *Rapp. Comm. Int. Mer Médit.* 37,-415.

Palanques, A., Durrieu de Madron, X., Puig, P., Fabres, J., Guillén, J., Calafat, A., Canals, M., Heussner, S., Bonnin, J., 2006. Suspended sediment fluxes and transport processes in the Gulf of Lions submarine canyons. The role of storms and dense water cascading. *Mar. Geol.* 234, 43-61.

Papaconstantinou, C., Kapiris K., 2003. Distribution and population structure of the red shrimp (*Aristeus antennatus*) on an unexploited fishing ground in the Greek Ionian Sea. *Aquat. Living Resour.* 14(5), 303-312.

Pardo, E., Aguilar, R., García, S., de la Torriente, A., Ubero, J., 2011. Documentación de arrecifes de corales de agua fría en el Mediterráneo occidental (Mar de Alborán). *Chronica naturae* 1, 20-34.

Parrish, F. A., Roark, E. B., 2009. Growth validation of gold coral *Gerardia* sp. in the Hawaiian Archipelago. *Mar. Ecol. Prog. Ser.* 397, 163–172.

Pérès, J. M., Piccard, J., 1964. Nouveau manuel de bionomie benthique de la mer Mediterranée. *Extrait du recueil des travaux de la station marine d'Endoume Bulletin 31,* Fasc. 47, 1-137.

Pérès, J. M., 1967. The Mediterranean benthos. *Oceanogr. Mar. Biol. Ann. Review* 5, 449-533.

Pérès, J. M., 1985. History of the Mediterranean biota and colonization of the depths. In: Margalef, R. (ed.) *Western Mediterranean.* Oxford: Pergamon Press, pp. 198-232.

Perkins, H., Pistek, P., 1990. Circulation in the Algerian Basin during June 1986. *J. Geophys. Res. Oceans* 95, 1577-1585.

Pinot, J. M., López-Jurado, J. L., Riera, M., 2002. The CANALES experiment (1996-1998). Interannual, seasonal and mesoscale variability of the circulation in the Balearic Channels. *Prog. Oceanogr.* 55, 335-370.

Politou, C.-Y., Karkani, M., Dokos, J., 2000. Distribution of Pandalid shrimps in Greek waters (Ionian Sea and Argosaronikos). In: M. Labropoulou (ed.), *Proceedings of the 6th Hellenic Symposium on Oceanography and Fisheries,* Chios, II, pp. 61-66. NCMR, Athens.

Politou, C.-Y., Kapiris, K., Maiorano, P., Capezzuto, F., Dokos, J., 2004. Deep-sea Mediterranean biology: the case of *Aristaeomorpha foliacea* (Risso, 1827) (Crustacea: Decapoda: Aristeidae). *Sci. Mar.* 68(3), 129-139.

Politou, C.-Y., Maiorano, P., D'Onghia, G., Mytilineou, C., 2005. Deep-water decapods crustacean fauna of the Eastern Ionian Sea. *Belg. J. Zool.* 135 (2): 235-241.

Psarra, S., Tselepides, A., Ignatiades, L., 2000. Primary productivity in the oligotrophic Cretan Sea (NE Mediterranean): seasonal and interannual variability. *Prog. Oceanogr.* 46, 187-204.

Puig, P., Company, J. B., Sardà, F., Palanqués, A., 2001. Responses of deep-water shrimp populations to intermediate nepheloid layer detachments on the Northwestern Mediterranean continental margin. *Deep-Sea Res. Pt. I* 48(10), 2195-2207.

Purser, A., Larsson, A. I., Thomsen, L., van Oevelen, D., 2010. The influence of flow velocity and food concentration on *Lophelia pertusa* (Scleractinia) zooplankton capture rates. *J. Exp. Mar. Biol. Ecol.* 395, 55-62.

Purser, A., Orejas, C., Gori, A., Unnithan, V., Thomsen, L., 2013. Local variation in the distribution of benthic megafauna species associated with cold-water coral reefs on the Norwegian margin. *Cont. Shelf Res.* 54, 37-51.

Quetglas, A., Alemany, F., Carbonell, A., Merella, P., Sanchez, P., 1998. Some aspects of the biology of *Todarodes sagittatus* (Cephalopoda: Ommastrephidae) from the Balearic Sea (western Mediterranean). *Sci. Mar.* 62, 73-82.

Quetglas, A., Alemany, F., Carbonell, A., Merella, P., Sanchez, P., 1999. Diet of the European flying squid *Todarodes sagittatus* (Cephalopoda: Ommastrephidae) in the Balearic Sea (western Mediterranean). *J. Mar. Biol. Assoc. UK* 79, 479-486.

Quetglas, A., Carbonell, A., Sanchez, P., 2000. Demersal continental shelf and upper slope cephalopod assemblages from the Balearic Sea (north-western Mediterranean). Biological aspects of some deep-sea species. *Est. Coast. Shelf Sci.* 50, 739-749.

Quetglas, A., Gonzalez, M., Carbonell, A., Sanchez, P., 2001. Biology of the deep-sea octopus *Bathypolypus sponsalis* (Cephalopoda: Octopodidae) from the western Mediterranean Sea. *Mar. Biol.* 138, 785-792.

Quetglas, A., Gonzalez, M., Franco, I., 2005. Biology of the upper-slope cephalopod *Octopus salutii* from the western Mediterranean Sea. *Mar. Biol.* 146, 1131-1138.

Quetglas, A., Ordines, F., Gonzalez, M., Franco, I., 2009. Life history of the bathyal octopus *Pteroctopus tetracirrhus* (Mollusca, Cephalopoda) in the Mediterranean Sea. *Deep-Sea Res. Pt. I* 56, 1379-1390.

Quetglas, A., de Mesa, A., Ordines, F., Grau, A., 2010. Life history of the deep-sea cephalopod family Histioteuthidae in the western Mediterranean. *Deep-Sea Res. Pt. I* 57, 999-1008.

Quetglas, A., Guijarro, B., Ordines, F., Massuti, E., 2012. Stock boundaries for fisheries assessment and management in the Mediterranean: the Balearic Islands as a case study. *Sci. Mar.* 76, 17-28.

Quetglas, A., Ordines, F., Gonzálea, M., Zaragoza, N., Mallol, S., Valls, M., de Mesa, A., 2013a. Uncommon pelagic and deep-sea cephalopods in the

Mediterranean: new data and literature review. *Med. Mar. Sci.* 14(1), 69-85.

Quetglas, A., Valls, M., Ordines, F., de Mesa, A., Olivar, M. P., Keller, S., Massutí, E., 2013b. Structure and dynamics of cephalopod assemblages in the water column on shelf-break and slope grounds of the western Mediterranean. *J. Marine Syst.,* http://dx.doi.org/10.1016/j.jmarsys.2013.11.015.

Quignard, J. P., Tomasini, J. A., 2000. Mediterranean fish biodiversity. *Biol. Mar. Medit.* 7(3), 1-66.

Ramírez-Llodra, E., Company, J. B., Camps, M., Rotllant, G., 2007. Spatio-temporal variations in reproductive patterns and population structure of *Pasiphaea multidentata* (Decapoda: Caridea) in the Blanes canyon and adjacent margin, North-western Mediterranean Sea. *Mar. Ecol. Evol. Persp.* 28(4), 470-479.

Rex, M. A., 1981. Community structure in the deep-sea benthos. *Annu. Rev. Ecol. Syst.* 12, 331-353.

Rex, M. A., Crame, J. A., Stuart, C. T., Clarke, A., 2005. Large-scale biogeographic patterns in marine mollusks: A confluence of history and productivity? *Ecology* 86, 2288-2297.

Rex, M. A., Etter, R. J., Morris, J. S., Crouse, J., McClain, C. R., Johnson, N. A., Stuart, C. T., Deming, J. W., Thies, R., Avery, R., 2006. Global bathymetric patterns of standing stock and body size in the deep-sea benthos. *Mar. Ecol. Prog. Ser.* 317, 1-8.

Reyss, D., 1972-73. Les canyons sous-marins de la mer catalane le rech du cap et le rech lacaze-duthiers IV - étude synécologique des peuplements de macrofaune benthique. *Vie Milieu* 23, 101-142.

Roberts, J. M.,Wheeler, A., Freiwald, A., Cairns, S., 2009. Cold-Water Corals. *The Biology and Geology of Deep-Sea Coral Habitats.* Cambridge University Press, Cambridge. 352 pp.

Robinson, A. R., Golnaraghi, M., Lesile, W. G., Artegiani, A., Hecht, A., Lazzoni, E., Michelato, A., Sansone, E., Theocharis, A., Unluata, U. 1991. The Eastern Mediterranean general circulation: features, structure and variability. *Dynam. Atmos. Oceans* 15, 215–240.

Rossi, L., 1959. Le specie di *Eunicella* (Gorgonaria) del Golfo di Genova. *Res. Ligusticae* 118, 203–225.

Rotllant, G., Moranta, J., Massutí, E., Sardà, F., Morales-Nin, B., 2002. Reproductive biology of three gadiform fish species through the mediterranean deep-sea range (147-1850 m). *Sci. Mar.* 66, 157-166.

Rotllant, G., Ribes, E., Company, J. B., Durfort, M., 2005. The ovarian maturation cycle of the Norway lobster *Nephrops norvegicus* (Linnaeus, 1758) (Crustacea, Decapoda) from the western Mediterranean Sea. *Int. J. Inver. Rep. Dev.* 48(1-3): 161-169.

Rotllant, G., Abad, E., Sarda, F., Abalos, M., Company, J., Rivera, J., 2006. Dioxin compounds in the deep-sea rose shrimp *Aristeus antennatus* (Risso, 1816) throughout the Mediterranean Sea. *Deep-Sea Res. Pt. I* 53, 1895-1906.

Rueggeberg, A., Fietzke, J., Liebetrau, V., Eisenhauer, A., Dullo, W.-C., Freiwald, A., 2008. Stable strontium isotopes (delta Sr-88/86) in cold-water corals - A new proxy for reconstruction of intermediate ocean water temperatures. *Earth Planet. Sci. Letters* 269, 569-574.

Sabatini, A., Follesa, M. C., Locci, I., Pendugiu, A. A., Pesci, P., Cau, A., 2007. Assemblages in a submarine canyon: influence of depth and time. *Hydrobiologia* 580, 265-271.

Salman, A., Katagan, T., 2002. Lessepsian immigrant cephalopods of the Mediterranean Sea. In: Ozturk, B. and Basusta, N. (Eds.) *Workshop on Lessepsian migration proceedings.* Turkish Marine Research Foundation, 9, 71-74.

Salman, A., Katagan, T., Benli, H. A., 1997. Bottom trawl teuthofauna of the Aegean Sea. *Arch. Fish. Mar. Res.* 45, 183-196.

Salman, A., Katagan, T., von Boletzky, S., 1999. New cephalopod molluscs in the eastern Mediterranean: Previously unnoted species or recent migrants? *Vie Milieu* 49, 11-17.

Sánchez, P., 1986. Distribución batimétrica y abundancia de algunos cefalópodos del mar Catalán. *Inv. Pesq.* 50, 237-245.

Sánchez, P., Belcari, P., Sartor, P., 1998. Composition and spatial distribution of cephalopods in two north-western Mediterranean areas. *S. Afr. J. Mar. Sci.* 20, 17-24.

Sardà, F., Cartes, J. E., Norbis, W., 1994. Spatiotemporal structure of the deep-water shrimp *Aristeus antennatus* (Decapoda, Aristeidae) population in the Western Mediterranean. *Fish. B-NOAA* 92(3), 599-607.

Sardà, F., 1998. *Nephrops norvegicus* (L): Comparative biology and fishery in the Mediterranean Sea. Introduction, conclusions and recommendations. *Sci. Mar.* 62, 5-15.

Sardà, F., Company, J. B., Maynou, F., 2003. Deep-sea shrimp *Aristeus antennatus* Risso 1816 in the Catalan Sea, a review and perspectives. *J. Northw. Atl. Fish. Sci.* 31, 127-136.

Sarda, F., Calafat, A., Flexas, M. M., Tselepides, A., Canals, M., Espino, M., Tursi, A., 2004a. An introduction to Mediterranean deep-sea biology. *Sci. Mar.* 68, 7-38.

Sardà, F., D'Onghia, G., Politou, C. Y., Company, J., Maiorano, P., Kapiris, K., 2004b. Deep-sea distribution, biological and ecological aspects of *Aristeus antennatus* (Risso, 1816) in the western and central Mediterranean Sea. *Sci. Mar.* 68, 117-127.

Sartor, P., Belcari, P., Carbonell, A., Gonzales, M., Quetglas, A., Sanchez, P., 1998. The importance of cephalopods to trawl fisheries in the western Mediterranean. *S. Afr. J. Mar. Sci.* 20, 67-72.

Savini, A., Vertino, A., Marchese, F., Beuck, L., Freiwald, A., 2014. Mapping Cold-water coral habitats at different scales within the Northern Ionian Sea (Central Mediterranean): An assessment of coral coverage and associated vulnerability. *PLOS ONE* 9:e87108.

Sbrana, M., Viva, C., Belcari, P., 2006. Fishery of the deep-water rose shrimp *Parapenaeus longirostris* (Lucas, 1846) (Crustacea: Decapoda) in the northern tyrrhenian sea (Western mediterranean). *Hydrobiologia* 557, 135-144.

Schroeder, K., Ribotti, A., Borghini, M., Sorgente, R., Perilli, A., Gasparini, G., 2008. An extensive western Mediterranean deep water renewal between 2004 and 2006. *Geophys. Res. Lett.* 35 L18605, doi:10.1029/2008GL035146.

Schröder, K., Josey, A., Herrmann, M., Grignon, L., Gasparini, G. P., Bryden, H. L., 2010. Abrupt warming and salting of the Western Mediterranean Deep Water: Atmospheric forcings and lateral advection. *J. Geophys. Res.* 115, C08029, doi: 10.1029/2009JC005749.

Seibel, B. A., Thuesen, E. V., Childress, J. J., Gorodezky, L. A., 1997. Decline in pelagic cephalopod metabolism with habitat depth reflects differences in locomotory efficiency. *Biol. Bull.* 192, 262-278.

Seibel, B. A., Childress, J. J., 2000. Metabolism of benthic octopods (Cephalopoda) as a function of habitat depth and oxygen concentration. *Deep-Sea Res. Pt. I* 47, 1247-1260.

Sherwood, O. A., Risk, M. J., 2007. Deep-Sea Corals: New Insights to Paleoceanography. In: Hillaire-Marcel, C., deVernal, A. (Eds.), Proxies in Late Cenozoic Paleoceanography. *Developments in Marine Geology* 1, pp. 491–522.

Sherwood, O. A., Edinger, E. N., Guilderson, T. P., Ghaleb, B., Risk, M. J., Scott, D. B., 2008. Late Holocene radiocarbon variability in Northwest Atlantic slope waters. *Earth Planet. Sci. Lett.* 275, 146-153.

Stefanescu, C., Rucabado, J., Lloris, D., 1992. Depth-size trends in western Mediterranean demersal deep-sea fishes. *Mar. Ecol. Progr. Ser.* 81, 205-213.

Stefanescu, C., Lloris, D., Rucabado, J., 1993. Deep-sea fish assemblages in the Catalan Sea (western Mediterranean) below a depth of 1000 m. *Deep-Sea Res. Pt. I* 40, 695-707.

Storelli, M. M., Perrone, V. G., Marcotrigiano, G. O., 2007. Organochlorine contamination (PCBs and DDTs) in deep-sea fish from the Mediterranean sea. *Mar. Poll. Bull.* 54, 1968-1971.

Storelli, M. M., Perrone, V. G., 2010. Detection and quantitative analysis of organochlorine compounds (PCBs and DDTs) in deep sea fish liver from Mediterranean Sea. *Environ. Sci. Poll. R.* 17, 968-976.

Taupier-Letage, I. and the EGYPT/EGITTO Teams, 2007. New elements on the surface circulation in the eastern basin of the Mediterranean. *Rapp. Comm. Int. Mer. Medit.* 38, 204.

Taupier-Letage, I. 2008. On the use of thermal infrared images for circulation studies: applications to the eastern Mediterranean basin, In: *Remote sensing of the European Seas,* edited by: Barale, V. and Gade, M., Springer Netherlands, 153–164.

Taviani, M., 2002. The Mediterranean benthos from late Miocene up to present: ten million years of dramatic climatic and geologic vicissitudes. *Biol. Mar. Medit.* 9, 445-463.

Taviani, M., Remia, A., Corselli, C., Freiwald, A., Malinverno, E., Mastrototaro, F., Savini, A., Tursi, A., 2005a. First geo-marine survey of living cold-water *Lophelia* reefs in the Ionian Sea (Mediterranean basin). *Facies* 50, 409-417.

Taviani, M., Freiwald, A., Zibrowius, H., 2005b. Deep coral growth in the Mediterranean Sea: an overview. In: Freiwald, et al. (eds.) Cold-water corals and ecosystems. *Erlangen Earth Conference Series* pp: 137-156.

Tecchio, S., Ramírez-Llodra, E., Sardà, F., Company, J. B., 2011a. Biodiversity of deep-sea demersal megafauna in western and central Mediterranean basins. *Sci. Mar.* 75(2), 341-350.

Tecchio, S., Ramírez-Llodra, E., Sardà, F., Company, J. B., Palomera, I., Mechó, A., Pedrosa-Pàmies, R., Sanchez-Vidal, A., 2011b. Drivers of deep Mediterranean megabenthos communities along longitudinal and bathymetric gradients. *Mar. Ecol. Progr. Ser.* 439, 181-192.

Tecchio, S., 2013. Mediterranean Deep-sea Ecosystems: Biodiversity, functioning and vulnerability. Universitat de Barcelona, *Doctoral Thesis,* 147 pp.

Tecchio, S., Ramírez-Lodra, E., Aguzzi, J., Sánchez-Vidal, A., Flexas, M. M., Sardà, F., Company, J. B., 2013. Seasonal fluctuations of deep megabenthos: finding evidence of standing stock accumulation in a flux-rich continental slope. *Progr. Oceanogr.* 118, 188-198.

Templado, J.,García-Carrascosa, M., Baratech, L., Capaccioni, R., López Ibor, A., Silvestre, R., Masso, C., 1986. Estudio preliminar de la fauna asociada a los fondos coralígenos del mar de Alborán (SE de España). *Bol. Inst. Esp. de Oceanog.* 3(4), 93-104.

Thessalou-Legaki, M., 1989. Allometric growth of male secondary sexual characters and dioecy in *Parapandalus narval* (Decapoda: Pandalidae). *J. Crustacean Biol.* 9, 595–600.

Thresher, R. E., 2009. Environmental and compositional correlates of growth rate in deep-water bamboo corals (Gorgonacea; Isididae). *Mar. Ecol. Prog. Ser.* 397, 187–196.

Titschack, J., Freiwald, A., 2005. Growth, deposition, and facies of Pleistocene bathyal coral communities from Rhodes, Greece. In: Freiwald, et al. (eds.) Cold-water corals and ecosystems. *Erlangen Earth Conference Series* pp: 41-59.

Tittensor, D. P., Baco, A. R., Brewin, P. E., Clark, M. R., Consalvey, M., Hall-Spencer, J., Rowden, A. A., Schlacher, T., Stocks, K. I., Rogers, A. D., 2009. Predicting global habitat suitability for stony corals on seamounts. *J. Biogeogr.* 36, 1111-1128.

Tortonese, E., 1960. General remarks on the Mediterranean deepsea fishes. *Bull. Inst. Océanogr.* Monaco 1167, 1-14.

Tsounis, G., Orejas, C., Reynaud, S., Gili, J. M., Allemand, D., Ferrier-Pagès, C., 2010. Prey-capture rates in four Mediterranean cold water corals. *Mar. Ecol. Prog. Ser.* 398, 149-155.

Tudela, S., 2004. Ecosystem effects of fishing in the Mediterranean: an analysis of the major threats of fishing gear and practices to biodiversity and marine habitats. *Studies and Reviews. General Fisheries Commission for the Mediterranean* 74, Rome, FAO, 44 pp.

Tursi, A., Mastrototaro, F., Matarrese, A., Maiorano, P., D'onghia, G., 2004. Biodiversity of the white coral reefs in the Ionian Sea (Central Mediterranean). *Chem. Ecol.* 20, 107-116.

Tyler, P. A., Zibrowius, H., 1992. Submersible observations of the invertebrate fauna on the continental slope southwest of Ireland. *Oceanol. Acta* 15, 211-226.

Tyler, P. A., Campos-Creasy, L. S., Giles, L. A., 1994. Environmental control of quai-continuous and seasonal reproduction in deep-sea benthic invertebrates. In: Young, C. M., Eckelbarger, K. J. (ed.) *Reproduction, larval biology, and recruitment of the deep-sea benthos*. Columbia University Press, New York, p. 158–178.

Tyler, P. A., 2003. The peripheral deep seas. In: Tyler, P. A. (Ed.) *Ecosystems of the deep oceans*. Elsevier Science, Amsterdam, The Netherlands, 261-293.

Vafidis, D., Koukouras, A., Voultsiadou-Koukoura, E., 1994. Octocoral fauna of the Aegean Sea with a check list of the Mediterranean species: new information, faunal comparisons. *Ann. Inst. Océanogr.*70, 217-229.

Vafidis, D., Politou, C. Y., Carbonell, A., and Company, J. B., 2005. A review of the biology and fisheries of the genus *Plesionika* Bate, 1888 (Decapoda, Caridea, Pandalidae) in European waters. *Crustaceana* 78, 335-352.

Vafidis, D., Leontarakis, P. K., Dailianis, T., Kallianiotis, A., 2008. Population characteristics of four deep-water pandalid shrimps (Decapoda: Caridea) in the northern Aegean Sea (NE Mediterranean). *J. Nat. Hist.* 42(31-32), 2079-2093.

Van de Flierdt, T., Robinson, L. F., Adkins, J. F., 2010. Deep-sea corals aragonite as a recorder for the neodymium isotopic composition of seawater. *Geochim. Cosmochim. Acta* 74, 6014–6032.

Vanney, J. R., Gennesseaux, M., 1985. Mediterranean seafloor features: Overview and assessment. In: Stanley, D. J. and Wezel, F. C. (Eds.) *Geological evolution of the Mediterranean Basin*. Springer, New York, 3-32.

Vargas-Yánez, M., García, M. J., Salat, J., García-Martínez, M. C., Pascual, J., Moya, F., 2008. Warming trends and decadal variability in the Western Mediterranean shelf. *Global Planet. Change* 63, 177-184.

Vargas-Yañez, M., Zunino, P., Schroeder, K., López-Jurado, J. L., Plaza, F., Serra, M., 2012. Extreme Western Intermediate Water formation in winter 2010. *J. Marine Syst.* 105-108, 52-59.

Vertino, A., Savini, A., Rosso, A., Di Geronimo, I., Mastrototaro, F., Sanfilippo, R., Gay, G., Etiope, G., 2010. Benthic habitat characterization and distribution from two representative sites of the deep-water SML Coral Province (Mediterranean). *Deep-Sea Res. Pt. II* 57, 380-396.

Villanueva, R., 1992. Deep-sea cephalopods of the north-western Mediterranean: indications of up-slope ontogenic migration in two bathybenthic species. *J. Zool.* 227, 267-276.

Voss, G. L. 1967. The biology and bathymetric distribution of deep-sea cephalopods. *Stud. Trop. Oceanogr.* 5, 511-535.

Voss, G. L., 1988. The biogeography of the deep-sea Octopoda. *Malacologia* 29, 295-307.

Waller, R. G., Tyler, P. A., Gage, J. D., 2002. Reproductive ecology of the deep-sea scleractinian coral *Fungiacyathus marenzelleri* (Vaughan, 1906) in the northeast Atlantic Ocean. *Coral Reefs* 21, 325-33.

Waller, R. G., 2005. Deep-water Scleractinia (Cnidaria: Anthozoa): current knowledge of reproductive processes. In: Freiwald, et al. (eds.) Cold-water corals and ecosystems. *Erlangen Earth Conference Series* pp: 691-700.

Waller, R. G., Tyler, P. A., 2005. The reproductive biology of two deep-water, reef-building scleractinians from the NE Atlantic Ocean. *Coral Reefs* 24, 514-522.

Waller, R. G., Tyler, P. A., Smith, C. R., 2008. Fecundity and embryo development of three Antarctic deep-water scleractinians: *Flabellum thouarsii, F. curvatum* and *F. impensum*. *Deep-Sea Res. Pt. II* 55, 2527-2534.

Waller, R. G., Tyler, P. A., 2011. Reproductive patterns in two deep-water solitary corals from the north-east Atlantic-*Flabellum alabastrum* and *F. angulare* (Cnidaria: Anthozoa: Scleractinia). *J. Mar. Biol. Ass. UK* 91, 669-675.

Weinberg, S., 1976. Revision of the common Octocorallia of the Mediterranean circalittoral. I. Gorgonacea. *Beaufortia* 24, 63–104.

Wienberg, C., Hebbeln, D., Fink, H. G., Mienis, F., Dorschel, B., Vertino, A., Lopez Correa, M., Freiwald, A., 2009. Scleractinian cold-water corals in the Gulf of Cadiz-First clues about their spatial and temporal distribution. *Deep-Sea Res. Pt. I* 56, 1873-1893.

Wright, R., 1978. Benthic foraminiferal repopulation of the Mediterranean after the Messinian (Late Miocen) event. *Palaeogeogr. Palaeoclimat. Palaeoecol.* 29, 189-214.

WWF/UICN, 2004. *The Mediterranean deep-sea ecosystems: an overview of their diversity, structure, functioning and anthropogenic impacts, with a proposal for conservation*. IUCN, Málaga and WWF, Rome.

Young, C. M., 2003. Reproduction, development and life-history traits. In: Tyler, P. A. (Ed.) *Ecosystems of the deep oceans*. Elsevier Science, Amsterdam, The Netherlands, 381-426.

Zibrowius, H., 1980. Les scléractiniaires de la Méditerranée et de l'Atlantique nord-oriental. *Mem. Inst. Oceanogr.* (Monaco) 11, 1-284.

In: Deep Sea
Editor: Nigel T. Wilson

ISBN: 978-1-63321-637-2

Chapter 3

PROKARYOTES IN METAL DEPOSITS ON THE DEEP SEAFLOOR

Shingo Kato[1,2] ***and Akihiko Yamagishi***[2*]

[1]Japan Collection of Microorganisms, RIKEN BioResource Center, Tsukuba, Ibaraki, Japan

[2]Department of Molecular Biology, Tokyo University of Pharmacy and Life Science, Hachioji, Tokyo, Japan

ABSTRACT

Approximately 70% of the surface of Earth is the bottom of the ocean (seabed or seafloor), about 80% of the seafloor is the deep seafloor (at deeper than 200 m below the sea-surface). Metal deposits accumulated on the seafloor have been found in various areas. Sulfide deposits containing metals such as Fe, Cu and Zn, called chimneys or mounds, and iron-silica-rich deposits occur in deep-sea hydrothermal fields. Ferromanganese deposits rich in Co, Cu, Ni and Zn, in addition to Fe and Mn, are found on the deep seafloor of both hydrothermal and non-hydrothermal fields. Previous studies have revealed that diverse prokaryotes (Bacteria and Archaea) are present on and within the seafloor metal deposits. Some species of prokaryotes can utilize Fe, Mn and S in respiration and fix inorganic carbon producing organic carbon. Others can

[*] Corresponding footnote: Akihiko Yamagishi, Department of Molecular Biology, Tokyo University of Pharmacy and Life Science, 1432-1 Horinouchi, Hachioji, Tokyo 192-0392, Japan. Telephone: +81-426-76-7139; Fax: +81-426-76-7145; E-mail: yamagish @toyaku.ac.jp.

accelerate dissolution or precipitation of metals. Metals such as Fe and Zn are essential for growth of almost all living organisms. Considering the wide distribution of the metal deposits on the deep seafloor and the amount of metals concentrated in the deposits, the activities of prokaryotes in the metal deposits potentially contribute to elemental cycling and maintenance of ecosystems in the global ocean. In this chapter, the recent studies of diversity and composition of prokaryotic communities in a variety of metal deposits on the deep seafloor are reviewed. Furthermore, the reported gene sequence data of the prokaryotes were re-analyzed using the latest bioinformatics tools. This provides novel insights into the characteristics of prokaryotic communities on the deep seafloor metal deposits.

INTRODUCTION

Prokaryotes, i.e., unicellular microorganisms belonging to the domains *Bacteria* and *Archaea*, are widely distributed on Earth, e.g., soils, human skins, and the deep seafloor environments. Culture-independent molecular biological analyses have been rapidly developed in the early 1990's, and these analyses targeting 16S rRNA genes have indicated that there are numerous yet-uncultured prokaryotes. It has been suggested that 99% of the total number of species present on Earth are still uncultivated (Amann et al. 1995; Rappé & Giovannoni 2003; Yarza et al. 2008). Most of the yet-uncultured prokaryotes are likely to be difficult to cultivate in laboratory because of slow growth, extremely restricted growth conditions, requirement for supports from other prokaryotes, etc. On-going development of the next generation sequencing (e.g., pyrosequencing) technology will provide much more sequences in shorter time and lower cost than traditional Sanger sequencing techniques (Shendure & Ji 2008). The next generation sequencing analyses have revealed that there are unexpectedly diverse prokaryotes in various environments, including many "rare" species in the communities (e.g., Sogin et al. 2006). As numerous 16S rRNA gene data have accumulated, well-ordered databases, e.g., SILVA (Quast et al. 2013), Greengenes (Desantis et al. 2006) and Ribosomal Database Project (RDP)(Cole et al. 2009), and useful bioinformatics tools, e.g., QIIME (Caporaso et al. 2010b), mothur (Schloss et al. 2009) and ARB (Ludwig et al. 2004), have been developed. It helps us to deal with the vast data to get a bird's-eye view of the diversity, distribution and composition of prokaryotic communities on Earth.

Over half of the surface of Earth is the deep seafloor (at deeper than 200 m below the sea-surface). Metal deposits containing such as Fe, Cu, Zn and Mn have been found on the deep seafloor in various areas (Rona 2003). Metals such as Fe and Zn are essential for almost all organisms. In addition, some prokaryotes utilize metals as energy sources. For better understanding of global energy flux and elemental cycling on Earth, it is important to reveal the distribution, diversity, abundance, composition, activity and productivity of prokaryotes in metal deposits on the deep-sea floor. Furthermore, we potentially exploit the metal-utilizing prokaryotes for clean and economical deep-sea mining or bioremediation in future. Based on the development of the analytical and sampling techniques, microbiological investigations on the deep seafloor, in particular, hydrothermal vents and methane seeps have been increasing. However, most areas of the deep seafloor, in particular, hydrothermally inactive fields still remain to be investigated by microbiological analyses. Hence, it is most likely that our knowledge of prokaryotes on the deep-seafloor is still extremely poor.

SPATIAL DISTRIBUTION AND CHEMICAL COMPOSITION OF SEAFLOOR METAL DEPOSITS

Metal deposits on the seafloor (Figure 1) can be classified by the origin of the metals, i.e., hydrothermal or hydrogenous. Hydrothermal deposits include polymetallic sulfides and iron-rich oxides, and hydrogenous deposits include ferromanganese (hereafter, Fe-Mn) crusts and nodules (Hein et al. 2010). Considering the wide distribution of the metal deposits on the deep-seafloor and the estimated total amount of metals, the metal deposits, in particular, polymetallic sulfide deposits and ferromanganese crusts, are the focus of commercial attention as a potential of mineral resources (Hein et al. 2010; Hoagland et al. 2010; Hannington et al. 2011).

Polymetallic Sulfides

Polymetallic sulfide deposits are found on the plate boundaries of which total length is 89,000 km, consist of mid-ocean ridges (65%), submarine volcanic island-arcs (12%) and back-arc basins (22%) (Bird 2003; Hannington et al. 2011). The size and mineralogy of the deposits are controlled by spreading rates of ridges, crustal thicknesses, heat flow pattern, water depths, and/or magma compositions (Hannington et al. 2011). The sulfide deposits contain such as Fe, Cu, Zn, Pb, and in some cases, Au and Ag (Herzig & Hannington 1995). The weight of the deposits have been estimated to be 6 × 10^{11} kg or more, containing ~3 × 10^{10} kg of Cu and Zn (Hannington et al. 2011). The sulfide deposits are generated by reaction between metal-rich hydrothermal fluids and oxygenated cold seawater. Chimney- or mound-like structures are observed in marine hydrothermal fields. In some cases, the sulfide deposits remain for tens of thousands years (Lalou et al. 1985).

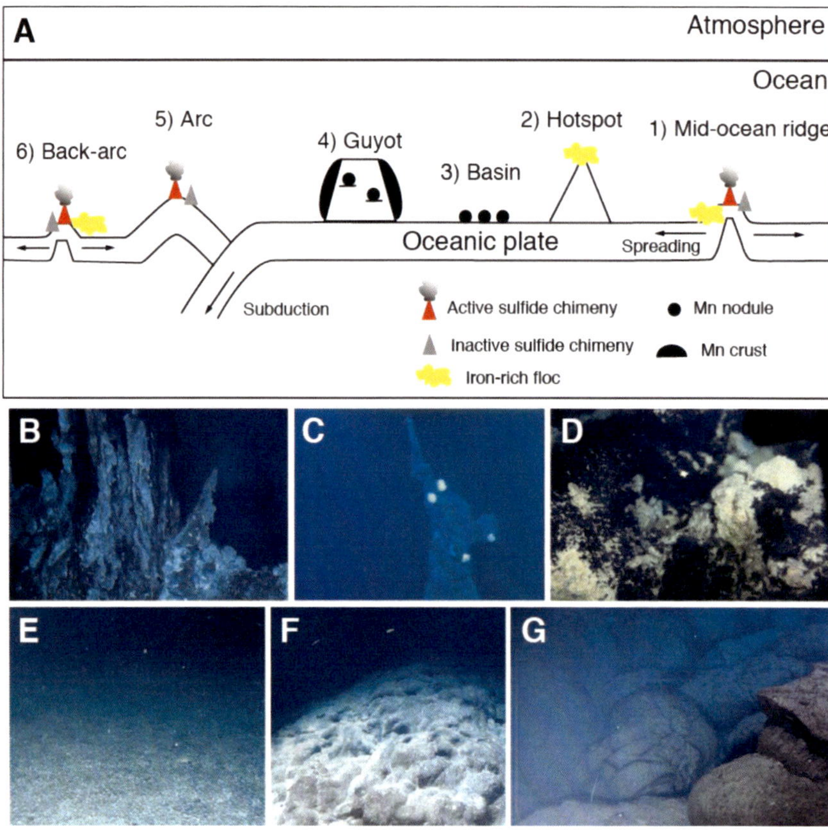

Figure 1. Metal deposits on deep seafloor. (A) A schematic illustration of distribution of metal deposits on variable geological settings. On-site photos of (B) active sulfide chimneys, (C) an inactive sulfide chimney and (D) iron-rich mats in the Southern Mariana Trough (SMT) back-arc hydrothermal field, (E) Mn nodules and (F) Mn crusts in Takuyo-Daigo seamount, and (G) basaltic pillow lavas in the SMT.

Iron-Rich Deposits

Iron-rich deposits have been found in deep-sea hydrothermal fields of various geological settings, such as mid-ocean ridges, e.g., Juan de Fuca Ridge (Davis et al. 2009), island-arcs, e.g., Tonga Arc (Langley et al. 2009a), back-arc basins, e.g., Southern Mariana Trough (Kato et al. 2009a), and hot spots, e.g., Loihi seamount (Emerson & Moyer 2002). The iron-rich deposits are often associated with relatively low temperature hydrothermal fluids (<100°C), which look like mats, flocs, or chimneys.

The precipitation of iron oxides is accompanied by activity of the iron-oxidizing bacteria (Emerson et al. 2010), which are called biogenic iron oxides (BIOS). BIOS contain extracellular polysaccharides (EPS) and accumulate metals on the surface of EPS and iron oxides (Ferris 2005; Langley et al. 2009b). Hence, iron-rich deposits consisting of BIOS may play a role in elemental cycling on marine hydrothermal systems.

Ferromanganese Deposits

Ferromanganese nodules and crusts (hereafter, Fe-Mn nodules and crusts, respectively) are widely distributed on the seafloor (Usui & Someya 1997; Hein et al. 2000; Rona 2003; Glasby 2006). The Fe-Mn nodules were discovered on the cruise of HMS Challenger during 1873-1876. Fe-Mn nodules are ball-shaped ferromanganese oxides found in deep-sea basins. Fe-Mn crusts are outcrops coated with ferromanganese oxides found on the sloping side of old seamounts. The growth rate of the Fe-Mn crusts is extremely slow (1-10 mm/Mya)(Hein et al. 2000; Usui et al. 2007).

Fe-Mn crusts and nodules mainly consist of Mn and Fe oxides (>30% of the total mass) as its name suggests (Mero 1962), and contain Co, Ni, Cu, Zn, rare earth elements and Pt (Hein et al. 2000). The Fe-Mn crusts contain a larger amount of these economically important elements ~10^4 times than Earth's crust (Hein et al. 2010). The amount of Fe-Mn oxides have been estimated based on the weight of Fe-Mn nodules in limited areas; for example, the total weight of manganese nodules in Clarion-Clipperton zone (defined between 0 and 30°N and 110 and 160°W) is estimated to be 6×10^{13} kg, containing 7×10^{12} kg Mn, 3×10^{11} kg each Ni and Cu, and 8×10^{10} kg Co (Morgan et al. 2000). In contrast, it is difficult to estimate the total amount of Fe-Mn crusts on the global seafloor because we don't have sufficient data regarding the thickness of ferromanganese-oxide layer of Fe-Mn crusts. An acoustic *in situ* measurement of the thickness of Fe-Mn layers reported recently help us with the more reliable estimation of the total amount of Fe-Mn crusts on the global seafloor (Thornton et al. 2010). The Fe-Mn nodules and crusts on the deep seafloor may play a significant role in elemental cycling in global oceans, because Fe-Mn oxides uptake elements (e.g., Ce, Co, Ni, Zn, in addition to Fe and Mn) from seawater (Koschinsky & Hein 2003).

Prokaryotic Communities in Seafloor Metal Deposits

Metals are essential for almost all living things, which is used mainly as a cofactor of enzymes. Some prokaryotic species oxidize or reduce Fe, Mn and S as electron donors or acceptors in their respiration. There are also some species that can accelerate dissolution or precipitation of metals. In contrast, high concentration of metals inhibits the growth of prokaryotes. Thus, metals and prokaryotes have strong relationships in nature. However the metal-prokaryote relationship in metal deposits on the deep seafloor has been poorly understood. Here, we summarize the recent studies of prokaryotic communities in a variety of metal deposits on the deep seafloor, i.e., active and inactive sulfides, Fe-Mn crusts, iron-rich mats and, in addition, basalts of Earth' crust.

Prokaryotic Abundance

The prokaryotic abundances in the deep seafloor metal deposits (10^5 – 10^{10} cells/g)(Table 1) are higher than those in deep-sea water (10^3 – 10^4 cells/ml)(Karner et al. 2001; Herndl et al. 2005). The cell counting results may be biased by the portion of the samples used (e.g., exterior or interior of the samples) and the counting methods e.g., direct cell counting with DAPI or SYBR green staining, fluorescence in situ hybridization (FISH) using bacterial and archaeal specific probes or quantitative PCR (Q-PCR). Overall, bacteria are dominated in the metal deposit samples. However, in some cases, over 10% (up to 65%) of the number of total cells were archaea. Therefore, archaea potentially play an important role in elemental cycling in deep-sea metal deposits. To better understand the whole microbial ecosystem, archaeal analysis should not be excluded.

Table 1. Abundance of prokaryotic cells and archaeal proportion

Sample types	Number of prokeyotic cells / g sample	Propotion of Archaea	Reference
Active sulfides	2.1×10^5 - 1.5×10^{10}	2.8% - 65%	Takai and Korikoshi, 1999; Takai et al., 2001; Schrenk et al., 2003; Kato et al., 2010
Iron-rich mats	2.0×10^8 - 3.8×10^8	12% - 18%	Kato et al., 2009a
Inactive sulfides	1.7×10^7 - 3.7×10^9	0.3% - 4.2%	Suzuki et al., 2004; Kato et al., 2010
Basalts	2.8×10^6 - 1.4×10^9	<0.1% - 12%	Einen eta l., 2004: Santelli et al., 2008
Mn crust	7.3×10^7	65%	Nitahara et al, 2011

Table 2. List of the clone libraries used in this chapter

Sample ID	Sample type	Location	Total # of clones	# of $OTUs_{0.05}$	Chao1*	Shannon score*	Original library ID	Reference
AAcs	Active chimeny	Southern Mariana Trough	543	136	27 (21-50)	2.68 (2.37-2.98)	AAcs	(Kato et al., 2010)
AFhm	Active chimeny	Southern Mariana Trough	59	15	17 (15-28)	2.02 (1.68-2.37)	AFhm	(Kato et al., 2010)
APbcs	Active chimeny	Southern Mariana Trough	167	71	69 (39-165)	3.05 (2.78-3.32)	APbcs	(Kato et al., 2010)
APcsc	Active chimeny	Southern Mariana Trough	119	68	88 (39-260)	2.69 (2.31-3.07)	APcsc	(Kato et al., 2010)
EPB2	Basaltic rock	East Pacific Rise	74	39	126 (70-280)	3.52 (3.30-3.74)	4059-B2(AlO)	(Santelli et al., 2008)
EPI2	Basaltic rock	East Pacific Rise	52	46	157 (91-320)	3.79 (3.60-3.99)	3965-I2(FeO1)	(Santelli et al., 2008)
EPMO	Basaltic rock	East Pacific Rise	84	59	225 (132-437)	4.04 (3.87-4.22)	3970-MO1A(MnO)	(Santelli et al., 2008)
EPO2	Basaltic rock	East Pacific Rise	79	45	305 (136-784)	3.56 (3.30-3.82)	3967-O2(FG)	(Santelli et al., 2008)
EPO8a	Basaltic rock	East Pacific Rise	81	54	348 (166-826)	3.92 (3.72-4.11)	3968-O8a(FeO2)	(Santelli et al., 2008)
P0X3b	Basaltic rock	Hawaii	64	43	102 (67-189)	3.60 (3.37-3.82)	PV550-X3(SP1)	(Santelli et al., 2008)
P0X4b	Basaltic rock	Hawaii	91	49	149 (89-298)	3.78 (3.59-3.97)	PV550-X4(SP2)	(Santelli et al., 2008)
P7X3b	Basaltic rock	Hawaii	71	25	152 (64-438)	2.13 (1.67-2.59)	PV547-X3(LSR)	(Santelli et al., 2008)

Table 2. (Continued)

Sample ID	Sample type	Location	Total # of clones	# of $OTUs_{0.05}$	Chao1*	Shannon score*	Original library ID	Reference
P9X2b	Basaltic rock	Hawaii	246	157	58 (29-159)	2.86 (2.55-3.17)	PV549-X2(LPP)	(Santelli et al., 2008)
3M23	Dead chimney	East Pacific Rise	77	41	147 (80-327)	3.52 (3.29-3.76)	3M23	(Sylvan et al., 2012)
3M33	Dead chimney	East Pacific Rise	72	34	129 (66-309)	3.11 (2.78-3.43)	3M33	(Sylvan et al., 2012)
7M23	Dead chimney	East Pacific Rise	65	27	53 (35-108)	2.89 (2.61-3.17)	7M24	(Sylvan et al., 2012)
9M32	Dead chimney	East Pacific Rise	65	30	44 (34-73)	3.14 (2.89-3.39)	9M32	(Sylvan et al., 2012)
9M4I	Dead chimney	East Pacific Rise	52	18	51 (26-149)	2.41 (2.14-2.68)	9M4I	(Sylvan et al., 2012)
9M4O	Dead chimney	East Pacific Rise	50	7	8 (7-15)	1.42 (1.17-1.67)	9M4O	(Sylvan et al., 2012)
9M4S	Dead chimney	East Pacific Rise	71	8	8 (8-14)	1.50 (1.20-1.79)	9M4S	(Sylvan et al., 2012)
IheB	Dead chimney	Okinawa Trough	174	11	13 (11-27)	2.01 (1.78-2.24)	Ihe	(Suzuki et al., 2004)
IndB	Dead chimney	Central Indian Ridge	242	20	31 (23-64)	2.55 (2.27-2.84)	Ind	(Suzuki et al., 2004)
INOC	Dead chimney	Southern Mariana Trough	92	39	112 (66-234)	3.38 (3.13-3.64)	INOC	(Kato et al., unpublished)
IPltc	Dead chimney	Southern Mariana Trough	271	133	140 (59-405)	3.00 (2.66-3.33)	IPltc	(Kato et al., 2010)
IYdc	Dead chimney	Southern Mariana Trough	144	53	19 (13-49)	1.97 (1.63-2.32)	IYdc	(Kato et al., 2010)
YS16	Iron-rich mat	Southern Mariana Trough	126	99	79 (45-178)	3.27 (3.02-3.52)	Y16	(Kato et al., 2009a)
YS18	Iron-rich mat	Southern Mariana Trough	125	89	206 (86-583)	3.00 (2.64-3.36)	Y18	(Kato et al., 2009a)
953Mn	Mn crust	Takuyo seamount	86	66	361 (191-760)	4.15 (3.98-4.32)	953Mn	(Nitahara et al., 2011)

* Normalized values to the lowest number of clones (50 for 9M4O) are shown. Numbers in parentheses indicate the 95% confidence interval.

Re-Analysis of Bacterial Community Diversity and Composition

Comparative analysis of the phylogenetic diversity and composition of the bacterial communities was performed with the published sequence data (Table 2). In this chapter, the clone library data containing 50 or more clones were used. The bacterial 16S rRNA gene sequences recovered from the metal deposits in the public databases (Table 2) were clustered into OTUs at 0.05 definition level (i.e., 95% similarity level) by uclust (Edgar 2010) in QIIME ver. 1.6.0 (Caporaso et al. 2010b). In general, $OTU_{0.03}$ or $OTU_{0.05}$ is used as species or genus level definition, respectively. The sequences deposited into public databases often contain only the representative sequences of each OTU, not all clones in the libraries. $OTU_{0.03}$ were usually used; however, $OTU_{0.01}$ or others were also used. It is known that the difference in the alignment and cluster methods used also affects the number of OTUs and the diversity values (Schloss 2010). Therefore, to compare diversity of the communities as fairly as possible, we added the replicated sequences of representative OTUs into the dataset to adjust the number of clones contained in each data set shown in the previous reports. $OTU_{0.05}$ clustering was performed using the modified dataset. The sequences in the dataset were aligned using PyNAST (Caporaso et al. 2010a). The quality of the alignment was manually checked in ARB (Ludwig et al. 2004). The taxonomic assignment for the clones was performed using the RDP classifier (Wang et al. 2007). The bacterial community structures in the samples are shown in Figure 2. The detailed detection frequency of detected clones in some of remarkable taxonomic groups in order level is shown in Table 3. Principal coordinate analysis (PCoA) was performed using UniFrac (Lozupone & Knight 2005; Lozupone et al. 2010) with unweighted algorithm and jackknifing replication. The PCoA plot with jackknifed support is shown in Figure 3. PCoA helps to visualize a few highly informative components of variation from multidimensional data. The above uclust, PyNAST, the RDP classifier and UniFrac were built-in programs of QIIME. Values of Chao1 richness estimators and Shannon diversity index were calculated by mothur (Schloss et al. 2009) with the distance matrix data generated by uclust and are shown in Table 2. Rarefaction curves were also generated by mothur. Rarefaction curves and Shannon diversity index values were summarized in Figures 4 and 5, respectively. The analysis for the archaeal communities was not performed because of insufficient data for comparative analysis.

Community Composition

Active Sulfide

Since the discovery of marine hydrothermal vents in the late 1970's, many studies have been done focusing on the prokaryotic communities in sulfide deposits actively venting hydrothermal fluids (Takai & Horikoshi 1999; Takai et al. 2001; Hoek et al. 2003; Schrenk et al. 2003; Alain et al. 2004; Kormas et al. 2006; Takai et al. 2008; Kato et al. 2010; Flores et al. 2011; Flores et al. 2012). One reason why the communities of active sulfides have studied is that microbial ecosystems in active vents are regarded as an analog of them in the early Earth (Nisbet & Sleep 2001; Reysenbach & Cady 2001; Kato & Yamagishi 2010). In fact, evidence for the presence of hyperthermophilic methanogens has obtained from Archaean hydrothermal fields (Ueno et al. 2006). Thermophilic archaea (e.g., *Thermoprotei*, *Archaeoglobi*, *Thermococci*, *Methanococci* and *Methanopyri*) and bacteria (e.g., *Aquificae*, *Thermodesulfobacteria* and *Thermotogae*) detected in the deep-sea active vents form a cluster in the deeper region in the phylogenetic tree of all life (Woese 1987; Stetter 2006). Interestingly, some deepest-branching uncultured archaea in the phylogenetic tree, such as AAG and MHVG1, have been detected in active vents (Takai & Horikoshi 1999; Kato et al. 2010); however, the physiologies of these archaea are still unclear.

In deep-sea active vents where sunlight does not reach, chemolithoautotrophs play important role as primary producers in the ecosystems. Chemolithoautotrophs gain energy by inorganic reactions between reduced species in hydrothermal fluids and oxidized species in seawater, and then using the energy they fix inorganic carbon forming organic carbon to construct their body. *Epsilonproteobacteria* is a major player in the communities in deep-sea active vents, which include sulfur- and hydrogen-oxidizing chemolithoautotrophs, mesophiles to moderate thermophiles, and aerobes to anaerobes (Takai et al. 2003; Nakagawa et al. 2005; Campbell et al. 2006). In fact, the relative abundance of *Epsilonproteobacteria* (in particular, *Campylobacterales*) was high in the active sulfides in the SMT (Figure 2 and Table 3). Although the detected memberships of prokaryotes are similar in various geological settings (Takai et al. 2006), the community structures (in particular, the relative abundance of chemolithoautotrophs) in active vents are strongly constrained by the fluid chemistry (McCollom & Shock 1997; Takai & Nakamura 2010; Amend et al. 2011)

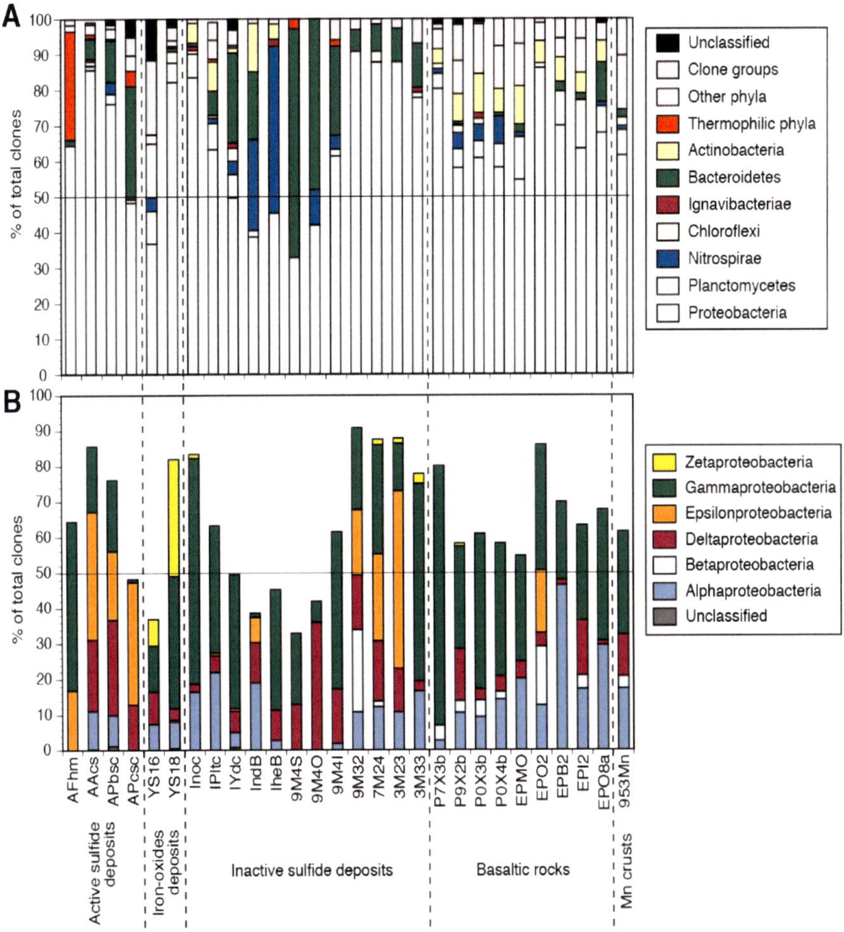

Figure 2. Bacterial community structures based on the 16S rRNA gene clone library analysis. Taxonomic distribution and relative abundance of each clone are shown in (A) phylum and (B) class level in Proteobacteria. The clone groups include BRC1, CD12, GN02, GN04, NC10, NKB19, NPL-UPA2, OD1, OP1, OP3, PAUC34f, SAR406, SBR1093, TM6, TM7, WS3 and ZB3. Other phyla include *Acidobacteria, "Caldithrix", Chlamydiae, Cyanobacteria, Firmicutes, Fusobacteria, Gemmatimonadetes, Lentisphaerae, Spirochaetes* and *Verrucomicrobia*. Thermophilic phyla include *Aquificae, Thermus-Deinococcus, Thermodesulfobacteria* and *Thermotogae*.

Inactive Sulfide

In contrast to active sulfides, there are few reports about the prokaryotes in sulfide deposits of inactive vents (Suzuki et al. 2004; Kato et al. 2010; Sylvan et al. 2012). Prokaryotes thriving in inactive sulfides are likely to be psychrophilic or psychrotolerant because the temperature in the inactive sulfide deposits must be similar to that of seawater (2-4°C). Several thermophiles and *Epsilonproteobacteria*, which are typical members in active sulfides, were detected in some inactive sulfides (Figure 2). They might be remnants when the vents were hydrothermally active. These inactive sulfide samples containing the thermophiles and *Epsilonproteobacteria* were grouped into the active sulfides (Figure 3). The energy sources for the prokaryotes in inactive sulfides may be reduced iron and sulfides and/or organic compounds that can be produced by chemolithoautotrophs while the vents were active. Oxygen should be depleted inside of the inactive sulfide deposits by microbial/chemical consumption on the surface. Therefore, it is expected that aerobes are present on the surface and anaerobes are present inside of the inactive sulfide deposits. In fact, OTUs affiliated with *Chromatiales* in *Gammaproteobacteria* and *Rhodobacterales* in *Alphaproteobacteria* including aerobic sulfide oxidizers, *Zetaproteobacteria* including microaerophilic iron oxidizers, *Deltaproteobacteria* including sulfate reducers, *Bacteroidetes* including fermenters and *Nitrospirae* including metal reducers were detected in the inactive sulfides, which were relatively abundant in some cases (Figure 2 and Table 3). In addition, OTUs affiliated with *Acidimicrobiales* in *Actinobacteria* and with *Gallionellales* in *Betaproteobacteria*, which contain iron oxidizers, were detected in some inactive sulfide samples with relatively high abundance in the libraries (up to 23.1%; Table 3).

The PCoA plot indicated that there are three different types among the inactive sulfide communities (Figure 3): The first type including 9M32, 7M24, 3M33 and 3M23 is clustered with the active sulfides and is relatively abundant in *Epsilonproteobacteria* (Figure 2); The second type including Inoc and IPltc is clustered with the basalts and is relatively abundant in *Gammaproteobacteria* (Figure 2); The third type including the others (i.e., IYdc, IheB, IndB, 9M4I, 9M4O and 9M4S) is clustered with only inactive sulfides and is relatively abundant in *Bacteroidetes* and *Nitrospirae* (Figure 2). The difference in the microbial community structures in inactive sulfides potentially reflects the elapsed time since the cease of hydrothermal activity as discussed previously (Kato et al. 2010; Sylvan et al. 2012).

Table 3. Detection frequency of detected clones in remarkable taxonomic groups in order level in terms of chemolithotrophy. The numbers indicate the percentages of each taxa in the number of the total clones

Taxon	Potential metabolisms	Active sulfides				Iron-rich mats			Inactive sulfides										
		Afhm	AAcs	APbsc	APcsc	YS16	YS18	Inoc	IPltc	IYdc	IndB	IheB	9M4S	9M4O	9M4I	9M32	7M24	3M23	3M33
Actinobacteria; Acidimicrobiia; Acidimicrobiales	Fe ox		0.4%	0.6%			0.5%	5.5%	8.1%	1.5%	13.7%	4.3%							
Proteobacteria; Alphaproteobacteria; Rhodobacterales	S ox, H_2 ox		6.5%	2.8%		1.2%	1.6%	1.1%	4.2%		9.5%					4.6%	3.1%	6.8%	9.7%
Proteobacteria; Betaproteobacteria; Gallionellales	Fe ox						0.5%									23.1%	1.5%		
Proteobacteria; Betaproteobacteria; Nitrosomonadales	NH_4 ox																		
Proteobacteria; Epsilonproteobacteria; Campylobacterales	H_2 ox, S ox	10.2%	35.8%	19.4%	27.6%				0.8%	0.7%	7.1%					18.5%	24.6%	50.0%	
Proteobacteria; Gammaproteobacteria; Chromatiales	S ox		7.9%	7.2%		0.6%	2.2%	41.8%	19.3%	4.4%	1.2%	12.1%			1.9%	13.8%	23.1%	2.7%	44.4%
Proteobacteria; Gammaproteobacteria; Methylococcales	CH_4 ox		1.5%			0.6%	4.9%		1.2%			5.0%						2.7%	2.8%

Table 3. (Continued)

	Basalts									Mn crust
Taxon	P7X3b	P9X2b	P0X3b	P0X4b	EPMO	EPO2	EPB2	EPI2	EPO8a	953Mn
Actinobacteria; Acidimicrobiia; Acidimicrobiales	4.2%	6.1%	10.9%	4.4%	9.5%	3.8%	6.8%	3.8%	6.2%	
Proteobacteria; Alphaproteobacteria; Rhodobacterales		0.4%					1.4%	1.9%	4.9%	
Proteobacteria; Betaproteobacteria; Gallionellales										
Proteobacteria; Betaproteobacteria; Nitrosomonadales		0.4%	1.6%	1.1%		1.3%				3.5%
Proteobacteria; Epsilonproteobacteria; Campylobacterales						17.7%				
Proteobacteria; Gammaproteobacteria; Chromatiales	1.4%	14.6%	3.1%	2.2%	4.8%	3.8%	5.5%	3.8%	3.7%	2.3%
Proteobacteria; Gammaproteobacteria; Methylococcales	2.8%	2.8%	14.1%	12.1%	10.7%	11.4%	1.4%		11.1%	4.7%

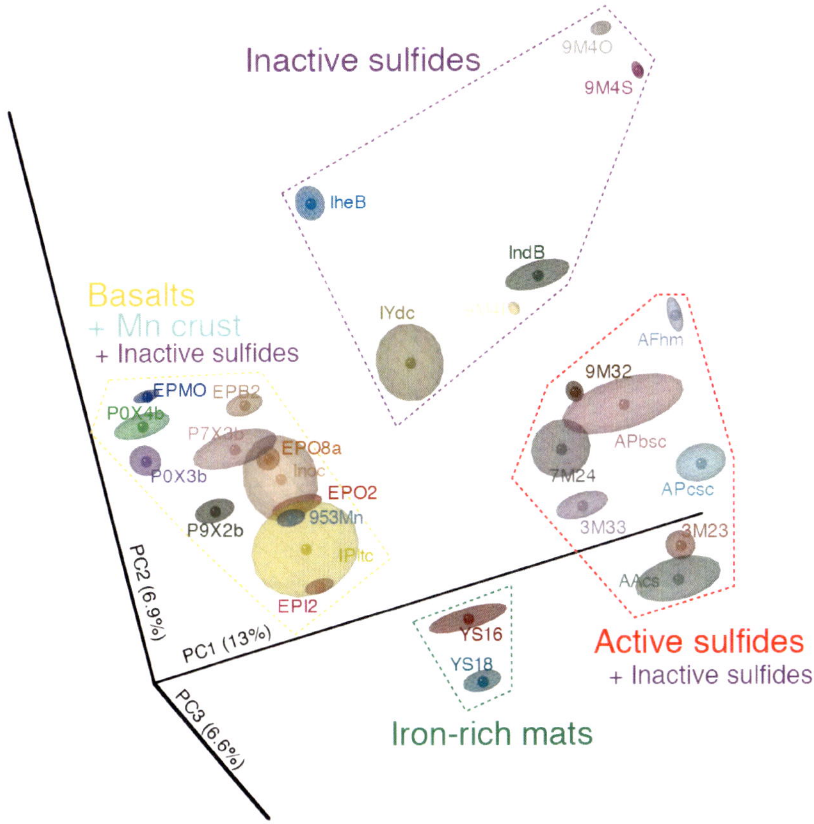

Figure 3. The result of PCoA with unweighted UniFrac, which accounts for presence/absence of taxa. The jackknife-supported (100 replicates) confidence ellipsoids are shown around the samples. The percentages in the axes represent the percentages of variation explained by first, second and third principal coordinates (PC1, 2 and 3).

In contrast to bacteria, only three taxonomic groups, i.e., Marine group I (MGI), Marine benthic group E (MBGE) and Deep-sea hydrothermal vent euryarchaeotic group 3 (DHVE3), were detected in the inactive sulfides (Suzuki et al. 2004; Kato et al. 2010). MGI contain ammonia oxidizers (Könneke et al. 2005) and are common members in seawater (Karner et al. 2001). Hence, it is difficult to exclude the possibility that they are contaminants from seawater. MBGE and DHVE3 are likely to be originated from the sulfides. It is noted that the bacteria and archaea detected in the inactive sulfide deposits by culture-independent analysis are mainly yet-uncultured species. Further cultivation efforts are needed to determine their physiological characteristics.

Iron-Rich Mat

The feature of iron-rich mat communities is the relatively high abundant of *Zetaproteobacteria* (Figure 2). This make the bacterial community structures in iron-rich mats distinguish from those in the other habitats (Kato et al. 2012b)(Figure 3). The zetaproteobacterial abundance of the number of total cells in the iron-rich mats ranged from 6% to 22%, as determined by quantitative PCR (Kato et al. 2009a). Members in *Zetaproteobacteria* were also abundant in the warm crustal fluids (up to 30ºC)(Kato et al. 2009b), suggesting that iron-rich deposits may be widely distributed in crustal aquifer. The appearance of microbial communities dominated by *Zetaproteobacteria* is likely to be related to the bioavailable energy yields by iron oxidation (Kato et al. 2012b). A variety of $OTUs_{0.03}$ (i.e., species level) have been detected in iron-rich mats in various areas (Kato et al. 2009a; Mcallister et al. 2011). However, only one species, autotrophic microaerophilic neutrophilic iron-oxidizing bacterium *Mariprofundus ferrooxidans*, has been reported (Emerson et al. 2007). It is still unclear whether all of the members in *Zetaproteobacteria* are iron oxidizers or not.

In the iron-rich mats, unique structures were observed, such as twisted-ribbon-like stalks or tube-like sheaths (Emerson & Moyer 2002; Kato et al. 2009a). Stalk or sheath structure produced by bacteria in *Gallionella* or *Leptothrix* have been observed in freshwater iron-rich environments (Emerson et al. 2010; Suzuki et al. 2011; Kato et al. 2012a). *M. ferrooxidans* produce such twisted stalks consisting of extracellular polysaccharides with iron oxides (Chan et al. 2011). Sheath-forming bacteria have not been identified yet in marine iron-rich mats. These unique structures can be recognized as biosignatures that provide evidence for the present of these bacteria in environments in the past time. In fact, stalk and sheath structures have been found in aged deposits (Hofmann & Farmer 2000).

The archaeal OTUs in MGI and pSL12 group that may contain ammonia oxidizers and ammonia monooxygenase genes (*pmoA*) were detected in the iron-rich mats (Kato et al. 2009a). Interestingly, some of the MGI OTUs and *pmoA* genes were different from those detected in seawaters, suggesting that there are indigenous ammonia oxidizers thriving in the iron-rich mats. Several uncultured archaeal groups (e.g. MBGE) were detected in the iron-rich mats (Kato et al. 2009a). However, their physiological characteristics are still unknown.

Fe-Mn Crust

The presence of microorganisms in/on Fe-Mn nodules has been paid attention and the biological influence on the growth of the metal oxides have been discussed since the middle of 20th century (Graham & Cooper 1959; Greenslate 1974). Prokaryotes on Fe-Mn nodules have been studies by culture-dependent methods since approximately 50 years ago, suggesting the prokaryotes associated with Mn precipitation and dissolution (Ehrlich 1963). Microscopic analyses have suggested that the presence of microbes on Fe-Mn nodules and crust (Wang et al. 2009a; Wang et al. 2009b). To date, Mn-oxidizing bacteria (not chemolithoautotrophs) have been isolated (Tebo et al. 2005). Fungi also affect the Mn precipitation (Miyata et al. 2006; Santelli et al. 2011). Mechanisms of the growth of Mn oxides associated with microbial activity have been discussed based on these facts (Wang & Müller 2009). However, the diversity, abundance and composition of prokaryotic communities on the Fe-Mn crusts and nodules had been poorly understood.

In 2011, the prokaryotic community diversity, abundance and composition on an aged Fe-Mn crust have been analyzed by culture-independent molecular analysis, indicating that the bacterial community composition was different from those of the surrounding sediment and seawater (Nitahara et al. 2011). The prokaryotes detected in the Fe-Mn crust were mainly affiliated in *Proteobacteria*, in particular, *Gammaproteobacteria* (Nitahara et al. 2011)(Figure 2). It is still unclear what are they doing there because most of the members were not similar to cultured species. The bacterial community structure of the Fe-Mn crust is similar to those in basalts (Kato & Yamagishi 2011)(Figure 3). Therefore, the environmental conditions of the habitats in basalt and in Fe-Mn crust for bacteria may be similar each other: low temperature (2-4°C) and low concentration of organic compounds (i.e., oligotrophic).

One of the energy sources for the prokaryotic communities on Fe-Mn crust may be ammonia. In fact, the OTUs related to ammonia-oxidizing bacteria (*Nitrosomonadales*) and archaea (MGI) were detected in the Fe-Mn crust (Nitahara et al. 2011)(Table 3). The *pmoA* genes were also detected in Fe-Mn crust (unpublished data), supporting the above notion.

Basalt

Basalt consists of silicate, contains iron and sulfur, and is a common igneous rock composing the oceanic crust. The energy potential involved in the oxidation of iron and sulfur contained in the basalts can be converted into primary biomass production of $\sim 10^{12}$ g carbon per year (Bach & Edwards

2003). Several studies using culture-independent analysis have been reported (Thorseth et al. 2001; Lysnes et al. 2004; Einen et al. 2006; Mason et al. 2007; Einen et al. 2008; Mason et al. 2008; Santelli et al. 2008; Santelli et al. 2009). The previous reports indicated that bacteria were dominated in the prokaryotic communities (Table 1), and that the bacterial communities mainly consist of *Proteobacteria* (>50% of the number of total clones), in particular, *Gammaproteobacteria* and *Alphaproteobacteria* (Figure 2). In addition, the OTUs affiliated in *Actinobacteria*, *Planctomycetes*, and some uncultured clone groups are relatively abundant (Figure 2). Notably, the OTUs affiliated in *Acidimicrobiales* of *Actinobacteria*, which contain iron oxidizers, were detected in all of the basalt samples with relatively high abundance in the libraries (3.8% to 10.9%) like the case of some inactive sulfides, suggesting that they use iron contained in the basalts as an energy source and play a role in basalt ecosystems as primary producers.

Most of the detected OTUs have been recognized as novel species (i.e., yet-uncultured species), which show 97% or lower similarity of 16S rRNA gene sequences to cultured species, as well as Fe-Mn crust. Most of the members in the bacterial communities between EPR and Hawaii were common despite the long mutual distance (>5000 km) (Santelli et al. 2008) as confirmed in the present PCoA analysis (Figure 3). It should be noted that sulfide- or iron-oxidizing chemolithoautotrophs have not been isolated in basalts. Hence, it is controversial whether sulfide and iron are the primary energy sources supporting the microbial communities on the basalts or not.

All of the detected archaea in basalts were affiliated with MGI (Mason et al. 2007). The OTUs in MGI in the basalts were phylogenetically distinguished from those in the seawater, suggesting that they are basalt-specific MGI archaea (Mason et al. 2007).

Ecological Significance and Economical Interests

Considering the wide distribution of the metal deposits on the deep seafloor and the amount of metals concentrated in the deposits, the activities of prokaryotes in the metal deposits potentially contribute to elemental cycling and maintenance of ecosystems in the global ocean. Prokaryotes that can be used for recycling of metal resources are obtained from the deep seafloor metal deposits. However, as described above, most of the prokaryotes in the deposits are yet-uncultured species. Further cultivation efforts focusing on the prokaryotes in the deep seafloor metal deposits are needed to know their physiological characteristics and ecological significance.

Findings in Re-classification of 16S rRNA Gene Sequences

The reclassification of the 16S rRNA gene sequences using newly developed tools helped us to find uncultured clone groups (e.g., NC10 and SBR1093; Figure 2) that had been classified as "unidentified" or "unclassified" bacteria in the previous reports. This expands our knowledge on the distribution of these groups in natural environments. The OTUs in SBR1093 were detected in some inactive sulfides, basalts and Fe-Mn crust (0.4% - 7.0%). This group was also detected in some basalt samples (Mason et al. 2007). Therefore, this group is relatively rare but potentially common in the low temperature solid habitats on the seafloor. The recent report showed that NC10 contain anaerobic methane oxidizers (Ettwig et al. 2010). The OTUs in NC10 were detected in two basalt samples (P0X3b and P0X4b). This implies that there are anaerobic methane oxidizers on the basalts. Thus, further cultivation of members in these uncultured clone group will provide novel insights into their ecological significance on the metal deposits.

Phylogenetic Diversity of Bacterial Communities

The phylogenetic diversity of bacterial communities in the metal deposits on the deep-sea floor varies (Table 2, Figs. 4 and 5). The Fe-Mn curst and most of the basalts showed higher diversity than the others (Shannon values are 3.5 to 4, the number of OTUs at the 50 clones are > 30). Most of the active sulfides and the iron-rich mats showed relatively high diversity (Shannon values are > 2.5, the number of OTUs at the 50 clones are > 30). Some inactive sulfides, which were collected from the exterior parts of the samples, also showed relatively high diversity (Shannon values are > 2.5, the number of OTUs at the 50 clones are > 25). Half of the inactive sulfides, in addition to one basaltic rock and one active sulfide, showed lower diversity (Shannon values are < 2.5, the number of OTUs at the 50 clones are < 25). The diversity between basalts and inactive sulfides is significantly difference ($p = <0.01$, t-test).

High Diversity in Basalts and Fe-Mn Curst

On the surface of basalts and Fe-Mn curst, biofilm and/or microbial colonies are often found as supported by microscopic observation (Einen et al. 2008; Santelli et al. 2008; Wang et al. 2009a). In these biofilm or colonies, steep gradient of pH can be formed in mm- or μm scale by microbial production of organic acids. Oxygen concentrations may have gradient from aerobic to anaerobic by microbial consumption. The variety of environmental conditions support the highly diverse bacterial communities on the surface of basalts and Fe-Mn curst. In fact, the relatively abundant taxonomic groups in the basalts and Fe-Mn crust, e.g., *Alphaproteobacteria*, *Gammaproteobacteria*, *Actinobacteria*, and *Planctomycetes* (Figure 2), contain variable species from aerobes to anaerobes, acidophiles to alkaliphiles, or chemoorganoheterotrophs using variable carbon compounds. It should be noted that basalts and Fe-Mn crust are too hard to be crushed by a hammer. Therefore, the genomic DNA was extracted from the rind of the rocky samples in the previous reports (Mason et al. 2008; Santelli et al. 2008; Nitahara et al. 2011). There is no evidence that microbes are present in the interior of the rocky samples. Actually, DNA extraction from the interior of the rocky samples has been unsuccessful (unpublished data).

Moderate Diversity in Active Sulfides and Iron-Mats

In the case of active sulfides and iron-mats, the energy sources supporting the microbial ecosystems are reduced inorganic chemical species, such as hydrogen, sulfide and iron, which are copiously supplied from hydrothermal fluids. Chemolithoautotrophs, e.g., hydrogen- or sulfide-oxidizers in *Epsilonproteobacteria* or iron-oxidizers in *Zetaproteobacteria*, are dominating as primary producers in the ecosystem in the habitats. In fact, these chemolithoautotrophs accounted for high percentage of total clones in the libraries (Figure 2). This may result in underestimation of the real diversity. Rare species in these habitats can be easily overlooked in the clone library analysis with the limited number of clones. The next generation sequencing analysis will facilitate the assessment of more reliable diversity in the chemolithoautotroph-dominated habitats. Temperature gradient (>100°C of hydrothermal fluids to 2-4°C of seawater) is formed in active sulfides and iron-mats, in addition to the gradient of pH and oxygen concentrations (Tivey & Mcduff 1990; Tivey et al. 1990). The OTUs related to thermophiles and mesophiles were detected in these samples (Figure 2). This may be related to the relatively high diversity in these samples.

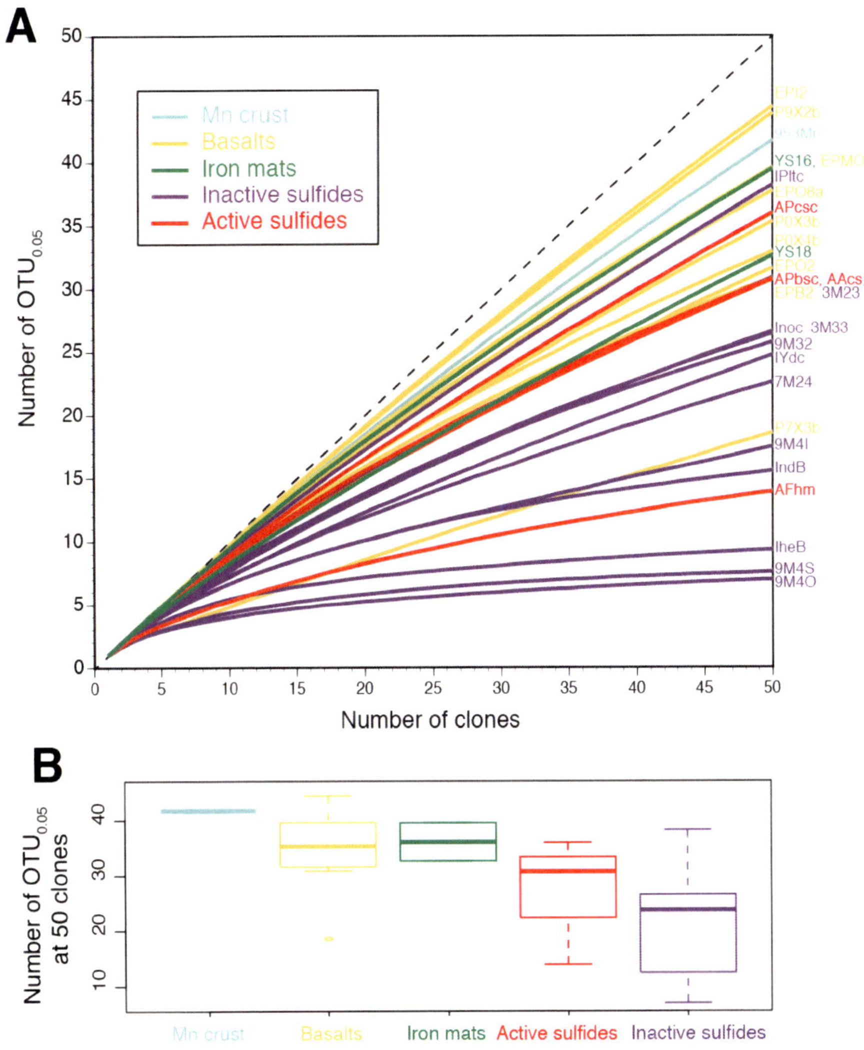

Figure 4. (A) Rarefaction curves for clone libraries of the deep seafloor metal deposit samples. (B) Box plot for comparison of the numbers of OTUs among the sample types. The numbers of OTU at 50 clones were used.

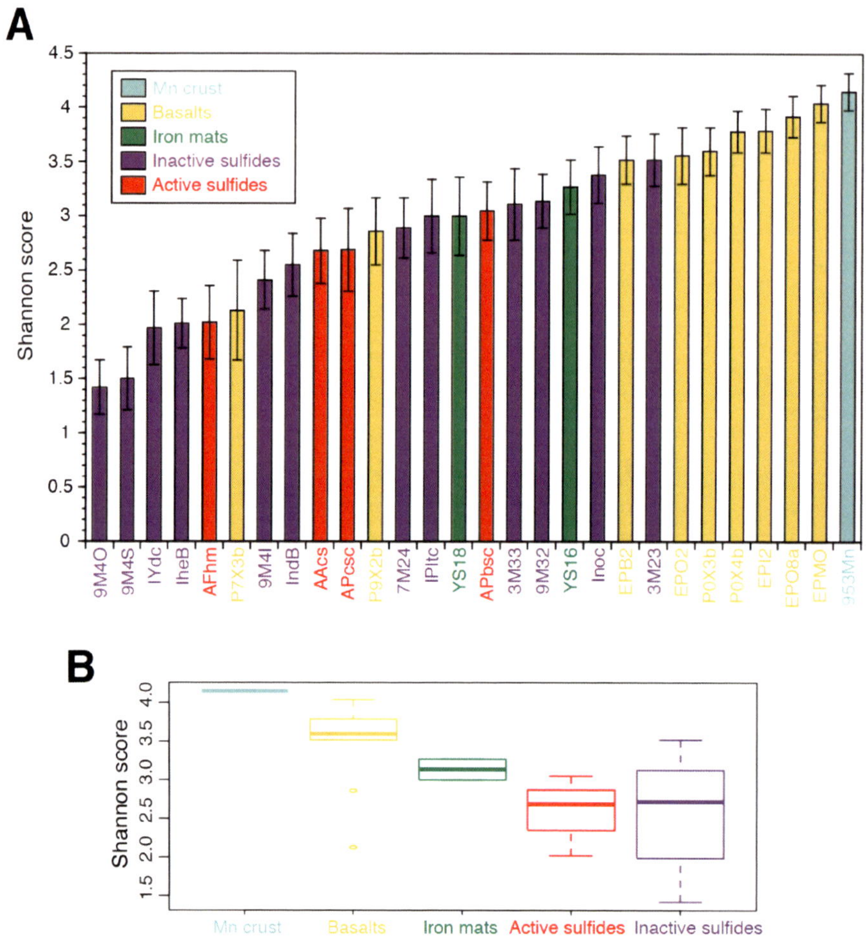

Figure 5. (A) Shannon diversity index values for clone libraries of the deep seafloor metal deposit samples. (B) Box plot for comparison of Shannon diversity index values among the sample types.

Low Diversity in Inactive Sulfides

In contrast, the bacterial communities in the inactive sulfides showed relatively low diversity (Figures 4 and 5). The low diversity of the bacterial communities in the inactive sulfides has been also observed from pyrosequencing analysis (Sylvan et al. 2012). This suggests that the environmental conditions in the inactive sulfides are not as varied as compared with those in active vents, iron-mats, and the surface portions of basalts and Fe-Mn crust. The oxygen and pH gradients may be formed in the surface of the inactive sulfides as well as basalts and Fe-Mn crust. However, the prokaryotes in the surface would be overlooked in the clone library analysis as described above. In contrast to the basalts and Fe-Mn crust, the inactive sulfides are fragile. Therefore, the DNA may be mainly originated from the interior parts of the sulfides, and less from the surface parts. The environmental conditions in the interior of the inactive sulfides are estimated to be the followings: the temperature is stable at 2-4°C as the same as the seawater; reduced iron and sulfide, and organic compounds produced by chemolithoautotrophs as described above are abundant as energy sources; the oxygen concentration is likely to be low or absent due to biotic/abiotic consumption. Specific members, such as *Rhodobacterales, Chromatiales, Acidimicrobiales, Nitrospirae, Bacteroidetes* and *Deltaproteobacteria* (Figure 2 and Table 3) can thrive in such conditions within the inactive sulfides.

At present, however, the affect on the observed diversity by methodological biases (e.g., the position and size of samples used, DNA extraction, the used thermal cycle conditions and primers in PCR and the used cloning methods) could not be ignored. Further comprehensive analysis using the same methods and careful sampling is needed to assess the relationship between the diversity and habitat types.

Conclusion

In this chapter, the previously reported microbial communities in the deep seafloor metal deposits were reanalyzed using recent bioinformatics tools such as QIIME and mother. The re-analysis of the communities in the metal deposit samples highlighted the characteristics of the composition and diversity in each habitat, e.g., active and inactive sulfides, iron-rich mats, Fe-Mn crust and basalts.

The re-analysis could be done within a day, if the sequence data would be ready to be analyzed. At present, the classification of the clones (or tags and reads for next generation sequencing analysis) using software is more impartial, accurate and rapid than by manual handling. The sequence data deposited into public databases (such as Genbank, EMBL and DDBJ) is accumulating continuously. The comparative analysis with the vast sequence data can be rapidly performed using a personal computer. The up-to-date bioinformatics tools will help us to look over the vast sequence data.

We realized again that the metal deposits are remarkable microbial and genetic resources as well as mineral resources. Although there are many yet-uncultured species, the number of cultured species will increase over time. More and more molecular investigations of deep seafloor metal deposits will be performed. Regular re-analysis with the latest sequence data and bioinformatics tools will provide novel insights into diversity, distribution and ecology of attractive prokaryotes on deep seafloor metal deposits.

ACKNOWLEDGMENTS

This research was funded by the Ministry of Education, Culture, Science and Technology (MEXT), Japan, through a special coordination fund (Project TAIGA: Trans-crustal Advection and In-situ biogeochemical processes of Global sub-seafloor Aquifer), by Grant-in-Aid for Young Scientists (B) from MEXT (no. 24770032), and partly by the RIKEN Special Postdoctoral Researchers Program.

REFERENCES

Alain K, Zbinden M, Bris NL, Lesongeur F, Quérellou J, Gaill F, Cambon-Bonavita M-A (2004) Early steps in microbial colonization processes at deep-sea hydrothermal vents. *Environ Microbiol* 6: 227-241.

Amann RI, Ludwig W, Schleifer KH (1995) Phylogenetic identification and in situ detection of individual microbial cells without cultivation. *Microbiol Rev* 59: 143-169.

Amend JP, McCollom TM, Hentscher M, Bach W (2011) Catabolic and anabolic energy for chemolithoautotrophs in deep-sea hydrothermal systems hosted in different rock types. *Geochimica et Cosmochimica Acta* 75: 5736-5748.

Bach W, Edwards KJ (2003) Iron and sulfide oxidation within the basaltic ocean crust: Implications for chemolithoautotrophic microbial biomass production. *Geochimica et Cosmochimica Acta* 67: 3871-3887.

Bird P (2003) An updated digital model of plate boundaries. *Geochemistry, Geophysics, Geosystems* 4: DOI: 10.1029/2001GC000252.

Campbell BJ, Engel AS, Porter ML, Takai K (2006) The versatile epsilon-proteobacteria: Key players in sulphidic habitats. *Nature Reviews: Microbiology* 4: 458-468.

Caporaso JG, Bittinger K, Bushman FD, DeSantis TZ, Andersen GL, Knight R (2010a) PyNAST: A flexible tool for aligning sequences to a template alignment. *Bioinformatics* 26: 266-267.

Caporaso JG, Kuczynski J, Stombaugh J, Bittinger K, Bushman FD, Costello EK, Fierer N, Pena AG, Goodrich JK, Gordon JI, Huttley GA, Kelley ST, Knights D, Koenig JE, Ley RE, Lozupone CA, McDonald D, Muegge BD, Pirrung M, Reeder J, Sevinsky JR, Turnbaugh PJ, Walters WA, Widmann J, Yatsunenko T, Zaneveld J, Knight R (2010b) QIIME allows analysis of high-throughput community sequencing data. *Nat Meth* 7: 335-336.

Chan CS, Fakra SC, Emerson D, Fleming EJ, Edwards KJ (2011) Lithotrophic iron-oxidizing bacteria produce organic stalks to control mineral growth: Implications for biosignature formation. *ISME J* 5: 717-727.

Cole JR, Wang Q, Cardenas E, Fish J, Chai B, Farris RJ, Kulam-Syed-Mohideen AS, McGarrell DM, Marsh T, Garrity GM, Tiedje JM (2009) The ribosomal database project: Improved alignments and new tools for rRNA analysis. *Nucleic Acids Research* 37: D141-D145.

Davis RE, Stakes DS, Wheat CG, Moyer CL (2009) Bacterial variability within an iron-silica-manganese-rich hydrothermal mound located off-axis at the cleft segment, Juan de Fuca Ridge. *Geomicrobiology Journal* 26: 570-580.

DeSantis TZ, Hugenholtz P, Larsen N, Rojas M, Brodie EL, Keller K, Huber T, Dalevi D, Hu P, Andersen GL (2006) Greengenes, a chimera-checked 16S rRNA gene database and workbench compatible with ARB. *Appl Environ Microbiol* 72: 5069-5072.

Edgar RC (2010) Search and clustering orders of magnitude faster than blast. *Bioinformatics* 26: 2460-2461.

Ehrlich HL (1963) Bacteriology of manganese nodules: I. Bacterial action on manganese in nodule enrichments. *Applied Microbiology* 11: 15-19.

Einen J, Thorseth IH, Øvreås L (2008) Enumeration of archaea and bacteria in seafloor basalt using real-time quantitative PCR and fluorescence microscopy. *FEMS Microbiology Letters* 282: 182-187.

Einen J, Kruber C, Øvreås L, Thorseth IH, Torsvik T (2006) Microbial colonization and alteration of basaltic glass. *Biogeosciences Discuss* 3: 273-307.

Emerson D, Moyer CL (2002) Neutrophilic Fe-oxidizing bacteria are abundant at the Loihi seamount hydrothermal vents and play a major role in Fe oxide deposition. *Appl Environ Microbiol* 68: 3085-3093.

Emerson D, Fleming EJ, McBeth JM (2010) Iron-oxidizing bacteria: An environmental and genomic perspective. *Annual Review of Microbiology* 64.

Emerson D, Rentz JA, Lilburn TG, Davis RE, Aldrich H, Chan C, Moyer CL (2007) A novel lineage of proteobacteria involved in formation of marine Fe-oxidizing microbial mat communities. *PLoS ONE* 2: e667. doi:610.1371/journal.pone.0000667.

Ettwig KF, Butler MK, Le Paslier D, Pelletier E, Mangenot S, Kuypers MMM, Schreiber F, Dutilh BE, Zedelius J, de Beer D, Gloerich J, Wessels HJCT, van Alen T, Luesken F, Wu ML, van de Pas-Schoonen KT, Op den Camp HJM, Janssen-Megens EM, Francoijs K-J, Stunnenberg H, Weissenbach J, Jetten MSM, Strous M (2010) Nitrite-driven anaerobic methane oxidation by oxygenic bacteria. *Nature* 464: 543-548.

Ferris FG (2005) Biogeochemical properties of bacteriogenic iron oxides. *Geomicrobiology Journal* 22: 79-85.

Flores GE, Shakya M, Meneghin J, Yang ZK, Seewald JS, Geoff Wheat C, Podar M, Reysenbach AL (2012) Inter-field variability in the microbial communities of hydrothermal vent deposits from a back-arc basin. *Geobiology* 10: 333-346.

Flores GE, Campbell JH, Kirshtein JD, Meneghin J, Podar M, Steinberg JI, Seewald JS, Tivey MK, Voytek MA, Yang ZK, Reysenbach A-L (2011) Microbial community structure of hydrothermal deposits from geochemically different vent fields along the Mid-Atlantic Ridge. *Environmental Microbiology* 13: 2158-2171.

Glasby G (2006) Manganese: Predominant role of nodules and crusts. *Marine Geochemistry*: 371-427.

Graham JW, Cooper SC (1959) Biological origin of manganese-rich deposits of the sea floor. *Nature* 183: 1050-1051.

Greenslate J (1974) Microorganisms participate in the construction of manganese nodules. *Nature* 249: 181-183.

Hannington M, Jamieson J, Monecke T, Petersen S, Beaulieu S (2011) The abundance of seafloor massive sulfide deposits. *Geology* 39: 1155-1158.

Hein JR, Conrad TA, Staudigel H (2010) Seamount mineral deposits: A source of rare metals for high-technology industries. *Oceanography* 23: 184-189.

Hein JR, Koschinsky A, Bau M, Manheim FT, Kang J-K, Roberts L (2000) Cobalt-rich ferromanganese crusts in the Pacific. In *Handbook of marine mineral deposits*. Cronan, Ds (ed). Boca Raton: CRC Press, pp. 239– 279.

Herndl GJ, Reinthaler T, Teira E, van Aken H, Veth C, Pernthaler A, Pernthaler J (2005) Contribution of archaea to total prokaryotic production in the deep atlantic ocean. *Appl Environ Microbiol* 71: 2303-2309.

Herzig PM, Hannington MD (1995) Polymetallic massive sulfides at the modern seafloor a review. *Ore Geology Reviews* 10: 95-115.

Hoagland P, Beaulieu S, Tivey MA, Eggert RG, German C, Glowka L, Lin J (2010) Deep-sea mining of seafloor massive sulfides. *Marine Policy* 34: 728-732.

Hoek J, Banta A, Hubler F, Reysenbach A-L (2003) Microbial diversity of a sulphide spire located in the edmond deep-sea hydrothermal vent field on the central indian ridge. *Geobiology* 1: 119-127.

Hofmann BA, Farmer JD (2000) Filamentous fabrics in low-temperature mineral assemblages: Are they fossil biomarkers? Implications for the search for a subsurface fossil record on the early earth and mars. *Planetary and Space Science* 48: 1077-1086.

Karner MB, DeLong EF, Karl DM (2001) Archaeal dominance in the mesopelagic zone of the Pacific Ocean. *Nature* 409: 507-510.

Kato S, Yamagishi A (2010) Microbial ecosystem in marine hydrothermal system in the early and present earth. *Viva Origino* 38: 46-55.

Kato S, Yamagishi A (2011) Microbial biodiversity and biogeography on the deep seafloor. In *Changing diversity in changing environment.* Grillo, O, Venora, G (eds): InTech.

Kato S, Kobayashi C, Kakegawa T, Yamagishi A (2009a) Microbial communities in iron-silica-rich microbial mats at deep-sea hydrothermal fields of the Southern Mariana Trough. *Environmental Microbiology* 11: 2094-2111.

Kato S, Kikuchi S, Kashiwabara T, Takahashi Y, Suzuki K, Itoh T, Ohkuma M, Yamagishi A (2012a) Prokaryotic abundance and community composition in a freshwater iron-rich microbial mat at circumneutral pH. *Geomicrobiology Journal* 29: 896–905.

Kato S, Nakamura K, Toki T, Ishibashi J, Tsunogai U, Hirota A, Ohkuma M, Yamagishi A (2012b) Iron-based microbial ecosystem on and below the seafloor: A case study of hydrothermal fields of the Southern Mariana Trough. *Frontiers in Microbiology* 3.

Kato S, Takano Y, Kakegawa T, Oba H, Inoue K, Kobayashi C, Utsumi M, Marumo K, Kobayashi K, Ito Y, Ishibashi J, Yamagishi A (2010) Biogeography and biodiversity in sulfide structures of active and inactive vents at deep-sea hydrothermal fields of the Southern Mariana Trough. *Appl Environ Microbiol* 76: 2968-2979.

Kato S, Yanagawa K, Sunamura M, Takano Y, Ishibashi J-i, Kakegawa T, Utsumi M, Yamanaka T, Toki T, Noguchi T, Kobayashi K, Moroi A, Kimura H, Kawarabayasi Y, Marumo K, Urabe T, Yamagishi A (2009b) Abundance of *Zetaproteobacteria* within crustal fluids in back-arc hydrothermal fields of the Southern Mariana Trough. *Environmental Microbiology* 11: 3210-3222.

Könneke M, Bernhard AE, de la Torre JR, Walker CB, Waterbury JB, Stahl DA (2005) Isolation of an autotrophic ammonia-oxidizing marine archaeon. *Nature* 437: 543-546.

Kormas KA, Tivey MK, Von Damm K, Teske A (2006) Bacterial and archaeal phylotypes associated with distinct mineralogical layers of a white smoker spire from a deep-sea hydrothermal vent site (9°N, East Pacific Rise). *Environmental Microbiology* 8: 909-920.

Koschinsky A, Hein JR (2003) Uptake of elements from seawater by ferromanganese crusts: Solid-phase associations and seawater speciation. *Marine Geology* 198: 331-351.

Lalou C, Brichet E, Hekinian R (1985) Age dating of sulfide deposits from axial and off-axial structures on the East Pacific Rise near 12°50'N. *Earth and Planetary Science Letters* 75: 59-71.

Langley S, Igric P, Takahashi Y, Sakai Y, Fortin D, Hannington M, Schwarz-Schampera U (2009a) Preliminary characterization and biological reduction of putative biogenic iron oxides (BIOS) from the tonga-kermadec arc, southwest Pacific Ocean. *Geobiology* 7: 35-49.

Langley S, Gault AG, Ibrahim A, Takahashi Y, Renaud R, Fortin D, Clark ID, Ferris FG (2009b) Sorption of strontium onto bacteriogenic iron oxides. *Environmental Science & Technology* 43: 1008-1014.

Lozupone C, Knight R (2005) UniFrac: A new phylogenetic method for comparing microbial communities. *Appl Environ Microbiol* 71: 8228-8235.

Lozupone C, Lladser ME, Knights D, Stombaugh J, Knight R (2010) UniFrac: An effective distance metric for microbial community comparison. *ISME J*.

Ludwig W, Strunk O, Westram R, Richter L, Meier H, Yadhukumar, Buchner A, Lai T, Steppi S, Jobb G, Forster W, Brettske I, Gerber S, Ginhart AW, Gross O, Grumann S, Hermann S, Jost R, Konig A, Liss T, Lussmann R, May M, Nonhoff B, Reichel B, Strehlow R, Stamatakis A, Stuckmann N, Vilbig A, Lenke M, Ludwig T, Bode A, Schleifer K-H (2004) ARB: A software environment for sequence data. *Nucl Acids Res* 32: 1363-1371.

Lysnes K, Thorseth IH, Steinsbu BO, Øvreås L, Torsvik T, Pedersen RB (2004) Microbial community diversity in seafloor basalt from the arctic spreading ridges. *FEMS Microbiology Ecology* 50: 213-230.

Mason OU, Di Meo-Savoie CA, Van Nostrand JD, Zhou J, Fisk MR, Giovannoni SJ (2008) Prokaryotic diversity, distribution, and insights into their role in biogeochemical cycling in marine basalts. *ISME J* 3: 231-242.

Mason OU, Stingl U, Wilhelm LJ, Moeseneder MM, Di Meo-Savoie CA, Fisk MR, Giovannoni SJ (2007) The phylogeny of endolithic microbes associated with marine basalts. *Environmental Microbiology* 9: 2539-2550.

McAllister SM, Davis RE, McBeth JM, Tebo BM, Emerson D, Moyer CL (2011) Biodiversity and emerging biogeography of the neutrophilic iron-oxidizing *Zetaproteobacteria*. *Appl Environ Microbiol* 77: 5445-5457.

McCollom TM, Shock EL (1997) Geochemical constraints on chemolithoautotrophic metabolism by microorganisms in seafloor hydrothermal systems. *Geochimica et Cosmochimica Acta* 61: 4375-4391.

Mero JL (1962) Ocean-floor manganese nodules. *Economic Geology* 57: 747-767.

Miyata N, Tani Y, Maruo K, Tsuno H, Sakata M, Iwahori K (2006) Manganese(IV) oxide production by *Acremonium sp.* Strain KR21-2 and extracellular Mn(II) oxidase activity. *Applied and Environmental Microbiology* 72: 6467-6473.

Morgan CL, Cronan D, Kennish M, Lutz P (2000) Resource estimates of the Clarion-Clipperton manganese nodule deposits. *Handbook of marine mineral deposits*: 145-170.

Nakagawa S, Takai K, Inagaki F, Hirayama H, Nunoura T, Horikoshi K, Sako Y (2005) Distribution, phylogenetic diversity and physiological characteristics of epsilon-proteobacteria in a deep-sea hydrothermal field. *Environmental Microbiology* 7: 1619-1632.

Nisbet EG, Sleep NH (2001) The habitat and nature of early life. *Nature* 409: 1083-1091.

Nitahara S, Kato S, Urabe T, Usui A, Yamagishi A (2011) Molecular characterization of the microbial community in hydrogenetic ferromanganese crusts of the Takuyo-Daigo seamount, northwest Pacific. *FEMS Microbiology Letters* 321: 121-129.

Quast C, Pruesse E, Yilmaz P, Gerken J, Schweer T, Yarza P, Peplies Jr, Glöckner FO (2013) The SILVA ribosomal RNA gene database project: Improved data processing and web-based tools. *Nucleic Acids Research* 41: D590-D596.

Rappé MS, Giovannoni SJ (2003) The uncultured microbial majority. *Annual Review of Microbiology* 57: 369-394.

Reysenbach A-L, Cady SL (2001) Microbiology of ancient and modern hydrothermal systems. *Trends in Microbiology* 9: 79-86.

Rona PA (2003) Resources of the sea floor. *Science* 299: 673-674.

Santelli CM, Edgcomb VP, Bach W, Edwards KJ (2009) The diversity and abundance of bacteria inhabiting seafloor lavas positively correlate with rock alteration. *Environmental Microbiology* 11: 86-98.

Santelli CM, Webb SM, Dohnalkova AC, Hansel CM (2011) Diversity of Mn oxides produced by Mn(II)-oxidizing fungi. *Geochimica et Cosmochimica Acta* 75: 2762-2776.

Santelli CM, Orcutt BN, Banning E, Bach W, Moyer CL, Sogin ML, Staudigel H, Edwards KJ (2008) Abundance and diversity of microbial life in ocean crust. *Nature* 453: 653-656.

Schloss PD (2010) The effects of alignment quality, distance calculation method, sequence filtering, and region on the analysis of 16S rRNA gene-based studies. *PLoS Comput Biol* 6: e1000844.

Schloss PD, Westcott SL, Ryabin T, Hall JR, Hartmann M, Hollister EB, Lesniewski RA, Oakley BB, Parks DH, Robinson CJ, Sahl JW, Stres B, Thallinger GG, Van Horn DJ, Weber CF (2009) Introducing mothur: Open-source, platform-independent, community-supported software for describing and comparing microbial communities. *Appl Environ Microbiol* 75: 7537-7541.

Schrenk MO, Kelley DS, Delaney JR, Baross JA (2003) Incidence and diversity of microorganisms within the walls of an active deep-sea sulfide chimney. *Appl Environ Microbiol* 69: 3580-3592.

Shendure J, Ji H (2008) Next-generation DNA sequencing. *Nat Biotech* 26: 1135-1145.

Sogin ML, Morrison HG, Huber JA, Welch DM, Huse SM, Neal PR, Arrieta JM, Herndl GJ (2006) Microbial diversity in the deep sea and the underexplored "rare biosphere". *Proceedings of the National Academy of Sciences* 103: 12115-12120.

Stetter KO (2006) Hyperthermophiles in the history of life. *Philosophical Transactions of the Royal Society B: Biological Sciences* 361: 1837-1843.

Suzuki T, Hashimoto H, Matsumoto N, Furutani M, Kunoh H, Takada J (2011) Nanometer-scale visualization and structural analysis of the inorganic/organic hybrid structure of *Gallionella ferruginea* twisted stalks. *Appl Environ Microbiol* 77: 2877-2881.

Suzuki Y, Inagaki F, Takai K, Nealson KH, Horikoshi K (2004) Microbial diversity in inactive chimney structures from deep-sea hydrothermal systems. *Microbial Ecology* 47: 186-196.

Sylvan JB, Toner BM, Edwards KJ (2012) Life and death of deep-sea vents: Bacterial diversity and ecosystem succession on inactive hydrothermal sulfides. *mBio* 3: e00279-00211. doi:00210.01128/mBio.00279-00211.

Takai K, Horikoshi K (1999) Genetic diversity of archaea in deep-sea hydrothermal vent environments. *Genetics* 152: 1285-1297.

Takai K, Nakamura K (2010) Compositional, physiological and metabolic variability in microbial communities associated with geochemically diverse, deep-sea hydrothermal vent fluids. In *Geomicrobiology: Molecular and environmental perspective*. Barton, Ll, Mandl, M, Loy, A (eds): Springer Netherlands, pp. 251-283.

Takai K, Komatsu T, Inagaki F, Horikoshi K (2001) Distribution of archaea in a black smoker chimney structure. *Appl Environ Microbiol* 67: 3618.

Takai K, Nakagawa S, Reysenbach AL, Hoek J (2006) Microbial ecology of mid-ocean ridges and back-arc basins. *Geophysical monograph* 166: 185-213.

Takai K, Inagaki F, Nakagawa S, Hirayama H, Nunoura T, Sako Y, Nealson KH, Horikoshi K (2003) Isolation and phylogenetic diversity of members of previously uncultivated ε-proteobacteria in deep-sea hydrothermal fields. *FEMS Microbiology Letters* 218: 167-174.

Takai K, Nunoura T, Ishibashi J-i, Lupton J, Suzuki R, Hamasaki H, Ueno Y, Kawagucci S, Gamo T, Suzuki Y, Hirayama H, Horikoshi K (2008) Variability in the microbial communities and hydrothermal fluid chemistry at the newly discovered mariner hydrothermal field, southern Lau Basin. *J Geophys Res* 113: G02031.

Tebo BM, Johnson HA, McCarthy JK, Templeton AS (2005) Geomicrobiology of manganese(II) oxidation. *Trends in Microbiology* 13: 421-428.

Thornton B, Asada A, Ura T, Ohira K, Kirimura D (2010) The development of an acoustic probe to measure the thickness of ferro-manganese crusts. In *OCEANS 2010 IEEE-Sydney*: IEEE, pp. 1-9.

Thorseth IH, Torsvik T, Torsvik V, Daae FL, Pedersen K-SP, R. B. (2001) Diversity of life in ocean floor basalt. *Earth and Planetary Science Letters* 194: 31-37.

Tivey MK, McDuff RE (1990) Mineral precipitation in the walls of black smoker chimneys: A quantitative model of transport and chemical reaction. *Journal of Geophysical Research* 95: 12617-12637.

Tivey MK, Olson LO, Miller VW, Light RD (1990) Temperature measurements during initiation and growth of a black smoker chimney. *Nature* 346: 51-54.

Ueno Y, Yamada K, Yoshida N, Maruyama S, Isozaki Y (2006) Evidence from fluid inclusions for microbial methanogenesis in the early archaean era. *Nature* 440: 516-519.

Usui A, Someya M (1997) Distribution and composition of marine hydrogenetic and hydrothermal manganese deposits in the northwest Pacific. *Geological Society, London, Special Publications* 119: 177-198.

Usui A, Graham IJ, Ditchburn RG, Zondervan A, Shibasaki H, Hishida H (2007) Growth history and formation environments of ferromanganese deposits on the Philippine Sea plate, northwest Pacific Ocean. *Island Arc* 16: 420-430.

Wang Q, Garrity GM, Tiedje JM, Cole JR (2007) Naïve bayesian classifier for rapid assignment of rRNA sequences into the new bacterial taxonomy. *Applied and Environmental Microbiology* 73: 5261-5267.

Wang X, Müller WEG (2009) Marine biominerals: Perspectives and challenges for polymetallic nodules and crusts. *Trends in Biotechnology* 27: 375-383.

Wang X, Schröder HC, Wiens M, Schloßmacher U, Müller WEG (2009a) Manganese/polymetallic nodules: Micro-structural characterization of exolithobiontic- and endolithobiontic microbial biofilms by scanning electron microscopy. *Micron* 40: 350-358.

Wang X, Schloßmacher U, Natalio F, Schröder HC, Wolf SE, Tremel W, Müller WEG (2009b) Evidence for biogenic processes during formation of ferromanganese crusts from the Pacific Ocean: Implications of biologically induced mineralization. *Micron* 40: 526-535.

Woese C (1987) Bacterial evolution. *Microbiol Rev* 51: 221-271.

Yarza P, Richter M, Peplies J, Euzeby J, Amann R, Schleifer K-H, Ludwig W, Glöckner FO, Rosselló-Móra R (2008) The all-species living tree project: A 16S rRNA-based phylogenetic tree of all sequenced type strains. *Systematic and Applied Microbiology* 31: 241-250.

INDEX

D

E

F

L

M

N

S